The Airlords of Han, by Philip Francis Nowlan

The Project Gutenberg EBook of The Airlords of Han, by Philip Francis Nowlan This eBook is for the use of anyone anywhere at no cost and with almost no restrictions whatsoever. You may copy it, give it away or re-use it under the terms of the Project Gutenberg License included with this eBook or online at www.gutenberg.net

Title: The Airlords of Han

Author: Philip Francis Nowlan

Illustrator: Frank R. Paul

Release Date: May 11, 2008 [EBook #25438]

Language: English

Character set encoding: ASCII

*** START OF THIS PROJECT GUTENBERG EBOOK THE AIRLORDS OF HAN ***

Produced by Greg Weeks, Stephen Blundell and the Online Distributed Proofreading Team at http://www.pgdp.net

---

Transcriber's Note:

This etext was produced from *Amazing Stories Fact and Science Fiction* May 1962 and was first published in *Amazing Stories* March 1929. Extensive research did not uncover any evidence that the U.S. copyright on this publication was renewed. Minor spelling and typographical errors have been corrected without note. A table of contents has been provided below:

I. The Airlords Besieged II. The "Ground Ships" Threaten III. We "Sink" the "Ground Ships" IV. Han Electrono-Ray Science V. American Ultronic Science VI. An Unequal Duel VII. Captured! VIII. Hypnotic Torture IX. The Fall of Nu-Yok X. Life in Lo-Tan, the Magnificent XI. The Forest Men Attack XII. The Mysterious "Air Balls" XIII. Escape! XIV. The Destruction of Lo-Tan XV. The Counter-Attack XVI. Victory

---

[Illustration: A Classic Reprint from AMAZING STORIES, March, 1929]

The AIRLORDS of HAN

By PHILIP FRANCIS NOWLAN

Illustrated by FRANK R. PAUL

Copyright, 1927, by E. P. Co., Inc.

Introduction by Sam Moskowitz

Of all the stories selected for reprinting by AMAZING STORIES, *none have received the acclamation of* Armageddon-2419, *the first Buck Rogers story published in the April, 1961, 35th Anniversary issue. Readers were amazed by the terse writing, the superb knowledge of military tactics, the brilliant prophecies of rocket guns, walkie talkies, jet planes, infra-red ray sights and many other devices which have become part and parcel of modern warfare.*

The Airlords of Han, *sequel to* Armageddon-2419, *is in every meaning of the phrase, a command performance for the readers of* AMAZING STORIES. *While many other stories may have received greater publicity through the years and other authors achieved more meaningful fame, Philip Francis Nowlan's work has not been unknown to prime movers in the field. At the time of Nowlan's sudden death from a stroke in 1940, John W. Campbell, Jr., who read the "Buck" Rogers stories when they first appeared and was at that very time scheduling the last story the man was ever to write, said: "The quality of Nowlan's written science-fiction was certainly exceptionally high--even ten years ago, when the magazine science-fiction was only starting, the work Phil Nowlan did was of a grade that would have been acceptable and well rated against the much more highly evolved work of today. He had, then, developed one of the concepts that has only recently been generally recognized and used; the realization that the thought-patterns of the people of the future will necessarily be as different from the everyday thought-matrices of our present as their background must differ from ours."*

The Airlords of Han *will prove a delight to the readers. Hugo Gernsback when "blurbing" it in 1929 said: "Mr. Nowlan has quite outdone himself. In our humble opinion, the sequel is in many respects better than the original story." That statement, is, in my opinion, true. This is a marvelous story from many standpoints.*

* * * * *

First of all from the aspect of the storyteller's art, it is told with a directness, precise imagery and discipline of controlled imagination that has relatively limited competition among science-fiction authors living or dead. The obvious grasp of military tactics, so strikingly evidenced in the first story, is reaffirmed here.

Here, too, the sound basis of the scientific prophecy has few equals. The war of the future is fought with rockets that sport atomic warheads! No indirection in prophecy. No broad gesture that could be interpreted as the foregoing, but the very terms spelled out for the very purpose we use them today and this is a story printed in 1929, not one appearing in 1944 or 1945, when the events were nearly upon us.

Buttressing the inventions are the sound, military reasons for their use and their limitations. You will find that the United States initially was devastated after a war with the Bolsheviks. You will stand back aghast as psychoanalysts (called that) practice brainwashing for political and military reasons. Public television for personal communication, like the telephone (recently instituted to a limited degree in Russia) is standard practice. Germ warfare is effectively practiced. Radio-controlled rocket-powered television scanners send back pictures and words to the operator, not unlike what some of our earth satellites are doing today.

All of this is placed logically in a society whose social system, business structure and mores are carefully integrated with the scientific material.

It was no accident that Buck Rogers became an almost instantaneous success as a comic strip. Behind its creation was one of the most logical, scientifically planned "worlds of if" ever conceived. The re-presentation of these original Buck Rogers stories secures for Philip Francis Nowlan the important place he deserves as a shaper of modern science fiction.

CHAPTER I

The Airlords Besieged

In a previous record of my adventures in the early part of the Second War of Independence I explained how I, Anthony Rogers, was overcome by radioactive gases in an abandoned mine near Scranton in the year 1927, where I existed in a state of suspended animation for nearly five hundred years; and awakened to find that the America I knew had been crushed under the cruel tyranny of the Airlords of Han, fierce Mongolians, who, as scientists now contend, had in their blood a taint not of this earth, and who with science and resources far in advance of those of a United States, economically prostrate at the end of a long series of wars with a Bolshevik Europe, in the year 2270 A.D., had swept down from the skies in their great airships that rode "repeller rays" as a ball rides the stream of a fountain, and with their terrible "disintegrator rays" had destroyed more than four-fifths of the American race, and driven the other fifth to cover in the vast forests which grew up over the remains of the once mighty civilization of the United States.

I explained the part I played in the fall of the year 2419, when the rugged Americans, with science secretly developed to terrific efficiency in their forest fastness, turned fiercely and assumed the aggressive against a now effete Han population, which for generations had shut itself up in the fifteen great Mongolian cities of America, having abandoned cultivation of the soil and the operation of mines; for these Hans produced all they needed in the way of food, clothing, shelter and machinery through electrono-synthetic processes.

I explained how I was adopted into the Wyoming Gang, or clan, descendants of the original populations of Wilkes-Barre, Scranton and the Wyoming Valley in Pennsylvania; how quite by accident I stumbled upon a method of destroying Han aircraft by shooting explosive rockets, not directly at the heavily armored ships, but at the repeller ray columns, which automatically drew the rockets upward where they exploded in the generators of the aircraft; how the Wyomings threw the first thrill of terror into the Airlords by bringing an entire squadron crashing to earth; how a handful of us in a rocketship successfully

CHAPTER I

raided the Han city of Nu-Yok; and how by the application of military principles I remembered from the First World War, I was able to lead the Wyomings to victory over the Sinsings, a Hudson River tribe which had formed a traitorous alliance with the hereditary enemies and oppressors of the White Race in America.

* * * * *

By the Spring of 2420 A.D., a short six months after these events, the positions of the Yellow and the White Races in America had been reversed. The hunted were now the hunters. The Hans desperately were increasing the defenses of their fifteen cities, around each of which the American Gangs had drawn a widely deployed line of long-gunners; while nervous air convoys, closely bunched behind their protective screen of disintegrator beams, kept up sporadic and costly systems of transportation between the cities.

During this period our own campaign against the Hans of Nu-Yok was fairly typical of the development of the war throughout the country. Our force was composed of contingents from most of the Gangs of Pennsylvania, Jersey and New England. We encircled the city on a wide radius, our line running roughly from Staten Island to the forested site of the ancient city of Elizabeth, to First and Second Mountains just west of the ruins of Newark, Bloomfield and Montclair, thence northeasterly across the Hudson, and down to the Sound. On Long Island our line was pushed forward to the first slopes of the hills.

We had no more than four long-gunners to the square mile in our first line, but each of these was equal to a battery of heavy artillery such as I had known in the First World War. And when their fire was first concentrated on the Han city, they blew its outer walls and roof levels into a chaotic mass of wreckage before the nervous Yellow engineers could turn on the ring of generators which surrounded the city with a vertical film of disintegrator rays. Our explosive rockets could not penetrate this film, for it disintegrated them instantly and harmlessly, as it did all other material substance with the sole exception of "inertron," that synthetic element developed by the Americans from the sub-electronic and ultronic orders.

CHAPTER I

* * * * *

The continuous operation of the disintegrators destroyed the air and maintained a constant vacuum wherever they played, into which the surrounding air continuously rushed, naturally creating atmospheric disturbances after a time, which resulted in a local storm. This, however, ceased after a number of hours, when the flow of air toward the city became steady.

The Hans suffered severely from atmospheric conditions inside their city at first, but later rearranged their disintegrator ring in a system of overlapping films that left diagonal openings, through which the air rushed to them, and through which their ships emerged to scout our positions.

We shot down seven of their cruisers before they realized the folly of floating individually over our invisible line. Their beams traced paths of destruction like scars across the countryside, but caught less than half a dozen of our gunners all told, for it takes a lot of time to sweep every square foot of a square mile with a beam whose cross section is not more than twenty or twenty-five feet in diameter. Our gunners, completely concealed beneath the foliage of the forest, with weapons which did not reveal their position, as did the flashes and detonation of the Twentieth Century artillery, hit their repeller rays with comparative ease.

The "drop ships," which the Hans next sent out, were harder to handle. Rising to immense heights behind the city's disintegrator wall, these tiny, projectile-like craft slipped through the rifts in the cylinder of destruction, and then turning off their repeller rays, dropped at terrific speed until their small vanes were sufficient to support them as they volplaned in great circles, shooting back into the city defenses at a lower level.

The great speed of these craft made it almost impossible to register a direct hit against them with rocket guns, and they had no repeller rays at which we might shoot while they were over our lines.

CHAPTER I

But by the same token they were able to do little damage to us. So great was the speed of a drop ship, that the only way in which it could use a disintegrator ray was from a fixed generator in the nose of the structure, as it dropped in a straight line toward its target. But since they could not sight the widely deployed individual gunners in our line, their scouting was just as ineffective as our attempts were to shoot them down.

* * * * *

For more than a month the situation remained a deadlock, with the Hans locked up in their cities, while we mobilized gunners and supplies.

Had our stock of inertron been sufficiently great at this period, we could have ended the war quickly, with aircraft impervious to the "dis" ray. But the production of inertron is a painfully slow process, involving the building up of this weightless element from ultronic vibrations through the sub-electronic, electronic and atomic states into molecular form. Our laboratories had barely begun production on a quantity basis, for we had just learned how to protect them from Han air raids, and it would be many months more before the supply they had just started to manufacture would be finished. In the meantime we had enough for a few aircraft, for jumping belts and a small amount of armor.

We Wyomings possessed one swooper completely sheathed with inertron and counterweighted with ultron. The Altoonas and the Lycomings also had one apiece. But a shielded swooper, while impervious to the "dis" ray, was helpless against squadrons of Han aircraft, for the Hans developed a technique of playing their beams underneath the swooper in such fashion as to suck it down flutteringly into the vacuum so created, until they brought it finally, and more or less violently, to earth.

Ultimately the Hans broke our blockade to a certain extent, when they resumed traffic between their cities in great convoys, protected by squadrons of cruisers in vertical formation, playing a continuous cross-fire of disintegrator beams ahead of them and down on the sides in a most effective screen, so that it was very difficult for us to get a rocket through to the repeller rays.

But we lined the scar paths beneath their air routes for miles at a stretch with concealed gunners, some of whom would sooner or later register hits, and it was seldom that a convoy made the trip between Nu-Yok and Bos-Tan, Bah-Flo, Si-ka-ga or Ah-la-nah without losing several of its ships.

Hans who reached the ground alive were never taken prisoner. Not even the splendid discipline of the Americans could curb the wild hate developed through centuries of dastardly oppression, and the Hans were mercilessly slaughtered, when they did not save us the trouble by committing suicide.

Several times the Hans drove "air wedges" over our lines in this vertical or "cloud bank" formation, ploughing a scar path a mile or more wide through our positions. But at worst, to us, this did not mean the loss of more than a dozen men and girls, and generally their raids cost them one or more ships. They cut paths of destruction across the map, but they could not cover the entire area, and when they had ploughed out over our lines, there was nothing left for them to do but to turn around and plough back to Nu-Yok. Our lines closed up again after each raid, and we continued to take heavy toll from convoys and raiding fleets. Finally they abandoned these tactics.

So at the time of which I speak, the Spring of 2420 A.D., the Americans and the Hans were temporarily at pretty much of a deadlock. But the Hans were as desperate as we were sanguine, for we had time on our side.

It was at this period that we first learned of the Airlords' determination, a very unpopular one with their conscripted populations, to carry the fight to us on the ground. The time had passed when command of the air meant victory. We had no visible cities nor massed bodies of men for them to destroy, nothing but vast stretches of silent forests and hills, where our forces lurked, invisible from the air.

CHAPTER II

The "Ground Ships" Threaten

One of our Wyoming girls, on contact guard near Pocono, blundered into a hunting camp of the Bad Bloods, one of the renegade American Gangs, which occupied the Blue Mountain section north of Delaware Water Gap. We had not invited their cooperation in this campaign, for they were under some suspicion of having trafficked with the Hans in past years, but they had offered no objection to our passage through their territory in our advance on Nu-Yok.

Fortunately our contact guard had been able to leap into the upper branches of a tree without being discovered by the Bad Bloods, for their discipline was lax and their guard careless. She overheard enough of the conversation of their Bosses around the camp fire beneath her to indicate the general nature of the Han plans.

After several hours she was able to leap away unobserved through the topmost branches of the trees, and after putting several miles between herself and their camp, she ultrophoned a full report to her Contact Boss back in the Wyoming Valley. My own Ultrophone Field Boss picked up the message and brought the graph record of it to me at once.

Her report was likewise picked up by the Bosses of the various Gang units in our line, and we had called a council to discuss our plans by word of mouth.

We were gathered in a sheltered glade on the eastern slope of First Mountain on a balmy night in May. Far to the east, across the forested slopes of the lowlands, the flat stretches of open meadow and the rocky ridge that once had been Jersey City, the iridescent glow of Nu-Yok's protecting film of annihilation shot upward, gradually fading into a starry sky.

In the faint glow of our ultronolamps, I made out the great figure and rugged features of Boss Casaman, commander of the Mifflin unit, and the gray uniform of Boss Warn, who led the Sandsnipers of the

CHAPTER II   11

Barnegat Beaches, and who had swooped over from his headquarters on Sandy Hook. By his side stood Boss Handan of the Winslows, a Gang from Central Jersee. In the group also were the leaders of the Altoonas, the Camerons, the Lycomings, Susquannas, Harshbargs, Hagersduns, Chesters, Reddings, Delawares, Elmirans, Kiugas, Hudsons and Connedigas.

* * * * *

Most of them were clad in forest-green uniforms that showed black at night, but each had some distinctive badge or item of uniform or equipment that distinguished his Gang.

Both the Mifflin and Altoona Bosses, for instance, wore heavy-looking boots with jointed knees. They came from sections that were not only mountainous, but rocky, where "leaping" involves many a slip and bruised limb, unless some protection of this sort is worn. But these boots were not as heavy as they looked, being counter-balanced somewhat with inertron.

The headgear of the Winslows was quite different from the close-fitting helmet of the Wyomings, being large and bushy-looking, for in the Winslow territory there were many stretches of nearly bare land, with occasional scrubby pines, and a Winslow caught in the open, on the approach of a Han airship, would twist himself into a motionless imitation of a scrubby plant, that passed very successfully for the real thing, when viewed from several thousand feet in the air.

The Susquannas had a unit that was equipped with inertron shields, that were of the same shape as those of the ancient Romans, but much larger, and capable of concealing their bearers from head to foot when they crouched slightly. These shields, of course, were colored forest green, and were irregularly shaded; they were balanced with inertron, so that their effective weight was only a few ounces. They were curious too, in that they had handles for both hands, and two small reservoir rocket-guns built into them as integral parts.

In going into action, the Susquannas crouched slightly, holding the shields before them with both hands, looking through a narrow vision slit, and working both rocket guns. The shields, however, were a great handicap in leaping, and in advancing through heavy forest growth.

The field unit of the Delawares was also heavily armored. It was one of the most efficient bodies of shock troops in our entire line. They carried circular shields, about three feet in diameter, with a vision slit and a small rocket gun. These shields were held at arm's length in the left hand on going into action. In the right hand was carried an ax-gun, an affair not unlike the battle-ax of the Middle Ages. It was about three feet long. The shaft consisted of a rocket gun, with an ax-blade near the muzzle, and a spike at the other end. It was a terrible weapon. Jointed leg-guards protected the ax-gunner below the rim of his shield, and a hemispherical helmet, the front section of which was of transparent ultron reaching down to the chin, completed his equipment.

* * * * *

The Susquannas also had a long-gun unit in the field.

One company of my Wyomings I had equipped with a weapon which I designed myself. It was a long-gun which I had adapted for bayonet tactics such as American troops used in the First World War, in the Twentieth Century. It was about the length of the ancient rifle, and was fitted with a short knife bayonet. The stock, however, was replaced by a narrow ax-blade and a spike. It had two hand-guards also. It was fired from the waist position.

In hand-to-hand work one lunged with the bayonet in a vicious, swinging up-thrust, following through with an up-thrust of the ax-blade as one rushed in on one's opponent, and then a down-thrust of the butt-spike, developing into a down-slice of the bayonet, and a final upward jerk of the bayonet at the throat and chin with a shortened grip on the barrel, which had been allowed to slide through the hands at the completion of the down-slice.

I almost regretted that we would not find ourselves opposed to the Delaware ax-men in this campaign, so curious was I to compare the efficiency of the two bodies.

But both the Delawares and my own men were elated at the news that the Hans intended to fight it out on the ground at last, and the prospect that we might in consequence come to close quarters with them.

Many of the Gang Bosses were dubious about our Wyoming policy of providing our fighters with no inertron armor as protection against the disintegrator ray of the Hans. Some of them even questioned the value of all weapons intended for hand-to-hand fighting.

As Warn, of the Sandsnipers put it: "You should be in a better position than anyone, Rogers, with your memories of the Twentieth Century, to appreciate that between the superdeadliness of the rocket gun and of the disintegrator ray there will never be any opportunity for hand-to-hand work. Long before the opposing forces could come to grips, one or the other will be wiped out."

But I only smiled, for I remembered how much of this same talk there was five centuries ago, and that it was even predicted in 1914 that no war could last more than six months.

* * * * *

That there would be hand-to-hand work before we were through, and in plenty, I was convinced, and so every able-bodied youth I could muster was enrolled in my infantry battalion and spent most of his time in vigorous bayonet practice. And for the same reason I had discarded the idea of armor. I felt it would be clumsy, and questioned its value. True, it was an absolute bar against the disintegrator ray, but of what use would that be if a Han ray found a crevice between overlapping plates, or if the ray was used to annihilate the very earth beneath the wearer's feet?

The only protective equipment that I thought was worth a whoop was a very peculiar device with which a contingent of five hundred Altoonas was supplied. They called it the "umbra-shield." It was a bell-shaped affair of inertron, counterweighted with ultron, about eight feet high. The gunner, who walked inside it, carried it easily with two shoulder straps. There were handles inside too, by which the gunner might more easily balance it when running, or lift it to clear any obstructions

CHAPTER II

on the ground.

In the apex of the affair, above his head, was a small turret, containing an automatic rocket gun. The periscopic gun sight and the controls were on a level with the operator's eyes. In going into action he could, after taking up his position, simply stoop until the rim of the umbra-shield rested on the ground, or else slip off the shoulder straps, and stand there, quite safe from the disintegrator ray, and work his gun.

But again, I could not see what was to prevent the Hans from slicing underneath it, instead of directly at it, with their rays.

* * * * *

As I saw it, any American who was unfortunate enough to get in the direct path of a "dis" ray, was almost certain to "go out," unless he was locked up tight in a complete shell of inertron, as for instance, in an inertron swooper. It seemed to me better to concentrate all our efforts on tactics of attack, trusting to our ability to get the Hans before they got us.

I had one other main unit besides my bayonet battalion, a long-gun contingent composed entirely of girls, as were my scout units and most of my auxiliary contingents. These youngsters had been devoting themselves to target practice for months, and had developed a fine technique of range-finding and the various other tactics of Twentieth Century massed artillery, to which was added the scientific perfection of the rocket guns and an average mental alertness that would have put the artilleryman of the First World War to shame.

From the information our contact guard had obtained, it appeared that the Hans had developed a type of "groundship" completely protected by a disintegrator ray "canopy" that was operated from a short mast, and spread down around it as a cone.

These ships were merely adaptations of their airships, and were designed to travel but a few feet above the ground. Their repeller rays were relatively weak; just strong enough to lift them about ten or twelve

CHAPTER II

feet from the surface. Hence they would draw but lightly upon the power broadcast from the city, and great numbers of them could be used. A special ray at the stern propelled them, and an extra-lift ray in the bow enabled them to nose up over ground obstacles. Their most formidable feature was the cone-shaped "canopy" of short-range disintegrator rays designed to spread down around them from a circular generator at the tip of a twenty-foot mast amidship. This would annihilate any projectile shot at it, for they naturally could not reach the ship without passing through the cone of rays.

It was instantly obvious that the "ground ships" would prove to be the "tanks" of the Twenty-fifth Century, and with due allowance for the fact that they were protected with a sheathing of annihilating rays instead of with steel, that they would have about the same handicaps and advantages as tanks, except that since they would float lightly on short repeller rays, they could hardly resort to the destructive crushing tactics of the tanks of the First World War.

* * * * *

As soon as our first supplies of inertron-sheathed rockets came through, their invulnerability would be at an end, as indeed would be that of the Han cities themselves. But these projectiles were not yet out of the factories.

In the meantime, however, the groundships would be hard to handle. Each of them we understood would be equipped with a thin long-range "dis" ray, mounted in a turret at the base of the mast.

We had no information as to the probable tactics of the Hans in the use of these ships. One sure method of destroying them would be to bury mines in their path, too deep for the penetration of their protecting canopy, which would not, our engineers estimated, cut deeper than about three feet a second. But we couldn't ring Nu-Yok with a continuous mine on a radius of from five to fifteen or twenty miles. Nor could we be certain beforehand of the direction of their attack.

In the end, after several hours' discussion, we agreed on a flexible defense. Rather than risk many lives, we would withdraw before them, test their effectiveness and familiarize ourselves with the tactics they adopted. If possible, we would send engineers in behind them from the flanks, to lay mines in the probable path of their return, providing their first attack proved to be a raid and not an advance to consolidate new positions.

CHAPTER III

We "Sink" the "Ground Ships"

Boss Handan, of the Winslows, a giant of a man, a two-fisted fighter and a leader of great sagacity, had been selected by the council as our Boss *pro tem*, and having given the scatter signal to the council, he retired to our general headquarters, which we had established on Second Mountain a few miles in the rear of the fighting front in a deep ravine.

There, in quarters cut far below the surface, he would observe every detail of the battle on the wonderful system of viewplates our ultrono engineers had constructed through a series of relays from ultroscope observation posts and individual "*cameramen*."

Two hours before dawn our long distance *scopemen* reported a squadron of "ground ships" leaving the enemy's disintegrator wall, and heading rapidly somewhat to the south of us, toward the site of the ancient city of Newark. The ultroscopes could detect no canopy operation. This in itself was not significant, for they were penetrating hills in their lines of vision, most of them, which of course blurred their pictures to a slight extent. But by now we had a well-equipped electronoscope division, with instruments nearly equal to those of the Hans themselves; and these could detect no evidence of *dis* rays in operation.

Handan appreciated our opportunity instantly, for no sooner had the import of the message on the Bosses' channel become clear than we heard his personal command snapped out over the long-gunners' general channel.

Nine hundred and seventy long-gunners on the south and west sides of the city, concealed in the dark fastnesses of the forests and hillsides, leaped to their guns, switched on their dial lights, and flipped the little lever combinations on their pieces that automatically registered them on the predetermined position of map section HM-243-839, setting their magazines for twenty shots, and pressing their fire buttons.

CHAPTER III

For what seemed an interminable instant nothing happened.

Then several miles to the southeast, an entire section of the country literally blew up, in a fiery eruption that shot a mile into the air. The concussion, when it reached me, was terrific. The light was blinding.

And our *scopemen* reported the instant annihilation of the squadron.

* * * * *

What happened, of course, was this; the Hans knew nothing of our ability to see at night through our ultroscopes. Regarding itself as invisible in the darkness, and believing our instruments would pick up its location when its *dis* rays went into operation, the squadron made the fatal error of not turning on its canopies.

To say that consternation overwhelmed the Han high command would be putting it mildly. Despite their use of code and other protective expedients, we picked up enough of their messages to know that the incident badly demoralized them.

Their next attempt was made in daylight. I was aloft in my swooper at the time, hanging motionless about a mile up. Below, the groundships looked like a number of oval lozenges gliding across a map, each surrounded by a circular halo of luminescence that was its *dis* ray canopy.

They had nosed up over the spiny ridge of what once had been Jersey City, and were moving across the meadowlands. There were twenty of them.

Coming to the darker green that marked the forest on the "map" below me, they adopted a wedge formation, and playing their pencil rays ahead of them, they began to beam a path for themselves through the forest. In my ears sounded the ultrophone instructions of my executives to the long-gunners in the forest, and one by one I heard the girls report their rapid retirement with their guns and other inertron-lightened equipment. I located several of them with my scopes, with which I could, of course, focus through the leafy screen

CHAPTER III

above them, and noted with satisfaction the unhurried speed of their movements.

On ploughed the Han wedge, while my girls separated before it and retired to the sides. With a rapidity much greater than that of the ships themselves, the beams penetrated deeper and deeper into the forest, playing continuously in the same direction, literally melting their way through, as a stream of hot water might melt its way through a snow bank.

Then a curious thing happened. One of the ships near one wing of the wedge must have passed over unusually soft ground, or perhaps some irregularity in the control of its canopy generator caused it to dig deeper into the earth ahead of it, for it gave a sudden downward lurch, and on coming up out of it, swerved a bit to one side, its offense beam slicing full into the ship echeloned to the left ahead of it. That ship, all but a few plates on one side, instantly vanished from sight. But the squadron could not stop. As soon as a ship stood still, its canopy ray playing continuously in one spot, the ground around it was annihilated to a continuously increasing depth. A couple of them tried it, but within a space of seconds, they had dug such deep holes around themselves that they had difficulty in climbing out. Their commanders, however, had the foresight to switch off their offense rays, and so damaged no more of their comrades.

* * * * *

I switched in with my ultrophone on Boss Handan's channel, intending to report my observation, but found that one of our swooper scouts, who, like myself, was hanging above the Hans, was ahead of me. Moreover, he was reporting a suddenly developed idea that resulted in the untimely end of the Hans' groundship threat.

"Those ships can't climb out of deep holes, Boss," he was saying excitedly. "Lay a big barrage against them--no, not on them--in front of them--always in front of them. Pull it back as they come on. But churn h--l out of the ground in front of them! Get the rocketmen to make a penetrative time rocket. Shoot it into the ground in front of them, deep enough to be below their canopy ray, see, and detonate under them as they go over it!"

I heard Handan's roar of exultation as I switched off again to order a barrage from my Wyoming girls. Then I threw my rocket motor to full speed and shot off a mile to one side, and higher, for I knew that soon there would be a boiling eruption below.

No smoke interfered with my view of it, for our atomic explosive was smokeless in its action. A line of blinding, flashing fire appeared in front of the groundship wedge. The ships ploughed with calm determination toward it, but it withdrew before them, not steadily, but jerkily intermittent, so that the ground became a series of gigantic humps, ridges and shell holes. Into these the Han ships wallowed, plunging ponderously yet not daring to stop while their protective canopy rays played, not daring to shut off these active rays.

One overturned. Our observers reported it. The result was a hail of rocket shells directly on the squadron. These could not penetrate the canopies of the other ships, but the one which had turned turtle was blown to fragments.

The squadron attempted to change its course and dodge the barrier in front of it. But a new barrier of blazing detonations and churned earth appeared on its flanks. In a matter of minutes it was ringed around, thanks to the skill of our fire control.

One by one the wallowing ships plunged into holes from which they could not extricate themselves. One by one their canopy rays were shut off, or the ships somersaulted off the knolls on which they perched, as their canopies melted the ground away from around them. So one by one they were destroyed.

Thus the second ground sortie of the Hans was annihilated.

CHAPTER IV

Han Electrono-Ray Science

At this period the Hans of Nu-Yok had only one airship equipped with their new armored repeller ray, their latest defense against our tactics of shooting rockets into the repeller rays and letting the latter hurl them up against the ships. They had developed a new steel alloy of tremendous strength, which passed their *rep* ray with ease, but was virtually impervious to our most powerful explosives. Their supplies of this alloy were limited, for it could be produced only in the Lo-Tan shops, for it was only there that they could develop the degree of electronic power necessary for its manufacture.

This ship shot out toward our lines just as the last of the groundships turned turtle and was blown to pieces. As it approached, the rockets of our invisible and widely scattered gunners in the forest below began to explode beneath its *rep* ray plates. The explosions caused the great ship to plunge and roll mightily, but otherwise did it no serious harm that I could see, for it was very heavily armored.

Occasionally rockets fired directly at the ship would find their mark and tear gashes in its side and bottom plates, but these hits were few. The ship was high in the air, and a far more difficult target than were its *rep* ray columns. To hit the latter, our gunners had only to gauge their aim vertically. Range could be practically ignored, since the *rep* ray at any point above two-thirds the distance from the earth to the ship would automatically hurl the rocket upward against the *rep* ray plate.

As the ship sped toward us, rocking, plunging and recovering, it began to beam the forest below. It was equipped with a superbeam too, which cut a swathe nearly a hundred feet wide wherever it played.

With visions of many a life snuffed out below me, I surrendered to the impulse to stage a single-handed attack on this ship, feeling quite secure in my floating shell of inertron. I nosed up vertically, and rocketed for a position above the ship. Then as I climbed upward, as yet unobserved in my tiny craft that was scarcely larger than myself, I trained my telultroscope on the Han ship, focussing through to a view

CHAPTER IV   22

of its interior.

Much as I had imbibed of this generation's hatred for the Hans, I was forced to admire them for the completeness and efficiency of this marvelous craft of theirs.

Constantly twirling the controls of my scope to hold the focus, I examined its interior from nose to stern.

* * * * *

It may be of interest at this point to give the reader a layman's explanation of the electronic or ionic machinery of these ships, and of their general construction, for today the general public knows little of the particular application of the electronic laws which the Hans used, although the practical application of ultronics are well understood.

Back in the Twentieth Century I had, like literally millions of others, dabbled a bit in "radio" as we called it then; the science of the Hans was simply the superdevelopment of "electricity," "radio," and "broadcasting."

It must be understood that this explanation of mine is not technically accurate, but only what might be termed an illustrative approximation.

The Hans' power-stations used to broadcast three distinct "*powers*" simultaneously. Our engineers called them the "*starter*," the "*pullee*" and the "*sub-disintegrator*." The last named had nothing to do with the operation of the ships, but was exclusively the powerizer of the disintegrator generators.

The "*starter*" was not unlike the "radio" broadcasts of the Twentieth Century. It went out at a frequency of about 1,000 kilocycles, had an amperage of approximately zero, but a voltage of two billion. Properly amplified by the use of *inductostatic* batteries (a development of the principle underlying the earth induction compass applied to the control of static) this current energized the *"A"* ionomagnetic coils on the airships, large and sturdy affairs, which operated the *Attractoreflex Receivers*, which in turn "pulled in" the second broadcast power known

CHAPTER IV   23

as the "*pullee*," absorbing it from every direction, literally exhausting it from surrounding space. The "*pullee*" came in at about a half-billion volts, but in very heavy amperage, proportional to the capacity of the receiver, and on a long wave--at audio frequency in fact. About half of this power reception ultimately actuated the *repeller ray* generators. The other half was used to energize the *"B" ionomagnetic* coils, peculiarly wound affairs, whose magnetic fields constituted the only means of insulating and controlling the circuits of the three "powers."

The repeller ray generators, operating on this current, and in conjunction with "twin synchronizers" in the power broadcast plant, developed two rhythmically variable ether-ground circuits of opposite polarity. In the "X" circuit, the negative was grounded along an ultraviolet beam from the ship's repeller-ray generator. The positive connection was through the ether to the "X synchronizer" in the power plant, whose opposite pole was grounded. The "Y" circuit travelled the same course, but in the opposite direction.

The rhythmic variables of these two opposing circuits, as nearly as I can understand it, in heterodyning, created a powerful material "push" from the earth, up along the violet ray beam against the *rep* ray generator and against the two synchronizers at the power plant.

This push developed molecularly from the earth-mass-resultant to the generator; and at the same fractional distance from the *rep* ray generator to the power plant.

* * * * *

The force exerted upward against the ship was, of course, highly concentrated, being confined to the path of the ultraviolet beam. Air or any material substance, coming within the indicated section of the beam, was thrown violently upward. The ships actually rode on columns of air thus forcefully up-thrown. Their "home berths" and "stations" were constructed with air pits beneath. When they rose from ordinary ground in open country, there was a vast upheaval of earth beneath their generators at the instant of take-off; this ceased as they got well above ground level.

CHAPTER IV

Equal pressure to the lifting power of the generator was exerted against the synchronizers at the power plant, but this force, not being concentrated directionally along an ultraviolet beam, involved a practical problem only at points relatively close to the synchronizers.

Of course the synchronizers were automatically controlled by the operation of the generators, and only the two were needed for any number of ships drawing power from the station, providing their protection was rugged enough to stand the strain.

Actually, they were isolated in vast spherical steel chambers with thick walls, so that nothing but air pressure would be hurled against them, and this, of course, would be self-neutralizing, coming as it did from all directions.

The "sub-disintegrator power" reached the ships as an ordinary broadcast reception at a negligible amperage, but from one to 500 "quints" (quintillions) voltage, controllable only by the fields of the *"B" ionomagnetic* coils. It had a wave-length of about ten meters. In the *dis* ray generator, this wave-length was broken up into an almost unbelievably high frequency, and became a directionally controlled wave of an infinitesimal fraction of an inch. This wave-length, actually identical with the diameter of an electron, that is to say, being accurately "tuned" to an electron, disrupted the orbital paths and balanced pulsations of the electrons within the atom, so desynchronizing them as to destroy polarity balance of the atom and causing it to cease to exist as an atom. It was in this way that the ray reduced matter to "nothingness."

This destruction of the atom, and a limited power for its reconstruction under certain conditions, marked the utmost progress of the Han science.

CHAPTER V

American Ultronic Science

Our own engineers, working in shielded laboratories far underground, had established such control over the "de-atomized" electrons as to dissect them in their turn into *sub-electrons*. Moreover, they had carried through the study of this "order" to the point where they finally "dissected" the *sub-electron* into its component *ultrons*, for the fundamental laws underlying these successive orders are not radically dissimilar. And as they progressed, they developed constructive as well as destructive practice. Hence the great triumphs of ultron and inertron, our two wonderful synthetic elements, built up from super-balanced and sub-balanced ultronic whorls, through the *sub-electronic* order into the *atomic* and *molecular*.

Hence also, come our relatively simple and beautifully efficient ultrophones and ultroscopes, which in their phonic and visual operation penetrate obstacles of material, electronic and sub-electronic nature without let or hindrance, and with the consumption of but infinitesimal power.

Static disturbance, I should explain, is negligible in the sub-electronic order, and non-existent in the ultronic.

The pioneer expeditions of our engineers into the ultronic order, I am told, necessitated the use of most elaborate, complicated and delicate apparatus, as well as the expenditure of most costly power, but once established there, all necessary power is developed very simply from tiny batteries composed of thin plates of *metultron* and *katultron*. These two substances, developed synthetically in much the same manner as ordinary ultron, exhibit dual phenomena which for sake of illustration I may compare with certain of the phenomena of radioactivity. As radium is constantly giving off electronic emanations and changing its atomic structure thereby, so *katultron* is constantly giving off *ultronic* emanations, and so changing its *sub-electronic* form, while *metultron*, its complement, is constantly attracting and absorbing *ultronic* values, and so changing its sub-electronic nature in the opposite direction. Thin plates of these two substances, when placed

CHAPTER V

properly in juxtaposition, with insulating plates of inertron between, constitute a battery which generates an ultronic current.

* * * * *

And it is a curious parallel that just as there were many mysteries connected with the nature of electricity in the Twentieth Century (mysteries which, I might mention, never *have* been solved, notwithstanding our penetration into the "sub-" orders) so there are certain mysteries about the ultronic current. It will flow, for instance, through an ultron wire, from the *katultron* to the *metultron* plate, as electricity will flow through a copper wire. It will short circuit between the two plates if the inertron insulation is imperfect. When the insulation is perfect, however, and no ultron metallic circuit is complete, the "current" (apparently the same that would flow through the metallic circuit) is projected into space in an absolutely straight line from the *katultron* plate, and received from space by the *metultron* plate on the same line. This line is the theoretical straight line passing through the mass-center of each plate. The shapes and angles of the plates have nothing to do with it, except that the perpendicular distance of the plate edges from the mass-center line determines thickness of the beam of parallel current-rays.

Thus a simple battery may be used either as a sender or receiver of current. Two batteries adjusted to the same center line become connected in series just as if they were connected by ultron wires.

In actual practice, however, two types of batteries are used; both the *foco* batteries and *broadcast* batteries.

Foco batteries are twin batteries, arranged to shoot a positive and a negative beam in the same direction. When these beams are made intermittent at light frequencies (though they are not light waves, nor of the same order as light waves) and are brought together, or focussed, at a given spot, the space in which they cross radiates alternating ultronic current in every direction. This radiated *ultralight* acts like true light so long as the crossing beams vibrate at light frequencies, except in three respects: first, it is not visible to the eye; second, its "color" is exclusively dependent on the frequency of the *foco* beams, which

CHAPTER V

determine the frequency of the alternating radiation. Material surfaces, it would appear, reflect them all in equal value, and the color of the resultant picture depends on the color of the *foco* frequencies. By altering these, a reddish, yellowish or bluish picture may be seen. In actual practice an orthochromatic mixture of frequencies is used to give a black, gray and white picture. The third difference is this: rays pulsating in line toward any ultron object connected with the rear plates of the twin batteries through rectifiers cannot be reflected by material objects, for it appears they are subject to a kind of "pull" which draws them straight through material objects, which in a sense are "magnetized" and while in this state offer no resistance.

Ultron, when so connected with battery terminals, glows with true light under the impact of *ultralight*, and if in the form of a lens or set of lenses, may be made to deliver a picture in any telescopic degree desired.

* * * * *

The essential parts of an ultroscope, then, are twin batteries with focal control and frequency control; an ultron shield, battery connected and adjustable, to intercept the direct rays from the "*glow-spot*," with an ordinary light-shield between it and the lens; and the lens itself, battery connected and with more or less telescopic elaboration.

To look through a substance at an object, one has only to focus the *glow-spot* beyond the substance but on the near side of the object and slightly above it.

A complete apparatus may be "set" for "penetrative," "distance" and "normal vision."

In the first, which one would use to look through the forest screen from the air, or in examining the interior of a Han ship or any opaque structure, the *glow-spot* is brought low, at only a tiny angle above the vision line, and the shield, of course, must be very carefully adjusted.

"Distance" setting would be used, for instance, in surveying a valley beyond a hill or mountain; the *glow-spot* is thrown high to illuminate the

CHAPTER V

entire scene.

In the "normal" setting the *foco* rays are brought together close overhead, and illuminate the scene just as a lamp of super brilliancy would in the same position.

For phonic communication a spherical sending battery is a ball of metultron, surrounded by an insulating shell of inertron, and this in turn by a spherical shell of katultron, from which the current radiates in every direction, tuning being accomplished by frequency of intermissions, with audio-frequency modulation. The receiving battery has a core pole of katultron and an outer shell of metultron. The receiving battery, of course, picks up all frequencies, the undesired ones being tuned out in detection.

Tuning, however, is only a convenience for privacy and elimination of interference in ultrophonic communication. It is not involved as a necessity, for untuned currents may be broadcast at voice-controlled frequencies, directly and without any carrier wave.

To use plate batteries or single center-line batteries for phonic communication would require absolutely accurate directional aligning of sender and receiver, a very great practical difficulty, except when sender and receiver are relatively close and mutually visible.

* * * * *

This, however, is the regular system used in the Inter-Gang network for official communication. The senders and receivers used in this system are set only with the greatest difficulty, and by the aid of the finest laboratory apparatus, but once set, they are permanently locked in position at the stations, and barring earthquakes or insecure foundations, need no subsequent adjustment. Accuracy of alignment permits beam paths no thicker than the old lead pencils I used to use in the Twentieth Century.

The non-interference of such communication lines, and the difficulty of cutting in on them from any point except immediately adjacent to the sender or receiver, is strikingly apparent when it is realized that every square inch of an imaginary plane bisecting the unlocated beam would have to be explored with a receiving battery in order to locate the beam itself.

A practical compromise between the spherical or universal broadcast senders and receivers on the one hand, and the single line batteries on the other, is the *multi-facet battery*. Another, and more practical device particularly for distance work, is the *window-spherical*. It is merely an ordinary spherical battery with a shielding shell with an opening of any desired size, from which a directionally controlled beam may be emitted in different forms, usually that simply of an expanding cone, with an angle of expansion sufficient to cover the desired territory at the desired point of reception.

CHAPTER VI

An Unequal Duel

But to return to my narrative, and my *swooper*, from which I was gazing at the interior of the Han ship.

This ship was not unlike the great dirigibles of the Twentieth Century in shape, except that it had no suspended control car nor gondolas, no propellers, and no rudders, aside from a permanently fixed double-fishtail stabilizer at the rear, and a number of "keels" so arranged as to make the most of the repeller ray airlift columns.

Its width was probably twice as great as its depth, and its length about twice its width. That is to say, it was about 100 feet from the main keel to the top-deck at their maximum distance from each other, about 200 feet wide amidship, and between 400 and 500 feet long. It had in addition to the top-deck, three interior decks. In its general curvature the ship was a compromise between a true streamline design and a flattened cylinder.

For a distance of probably 75 to 100 feet back of the nose there were no decks except that formed by the bottom of the hull. But from this point back the decks ran to within a few feet of the stern.

At various spots on the hull curvature in this great "hollow nose" were platforms from which the crews of the *dis* ray generators and the *electronoscope* and *electronophone* devices manipulated their apparatus.

Into this space from the forward end of the center deck, projected the control room. The walls, ceiling and floor of this compartment were simply the surfaces of *viewplates*. There were no windows or other openings.

The operation officers within the control room, so far as their vision was concerned, might have imagined themselves suspended in space, except for the transmitters, levers and other signalling devices around them.

CHAPTER VI

Five officers, I understand, had their posts in the control room; the captain, and the chiefs of *scopes*, *phones*, *dis rays* and *navigation*. Each of these was in continuous interphone communication with his subordinates in other posts throughout the ship. Each *viewplate* had its phone connecting with its "*eye machines*" on the hull, the crews of which would switch from telescopic to normal view at command.

There were, of course, many other *viewplates* at executive posts throughout the ship.

* * * * *

The Hans followed a peculiar system in the command of their ships. Each ship had a double complement of officers. Active Officers and Base Officers. The former were in actual, active charge of the ship and its apparatus. The latter remained at the ship base, at desks equipped with *viewplates* and phones, in constant communication with their "correspondents," on the ship. They acted continuously as consultants, observers, recorders and advisors during the flight or action. Although not primarily accountable for the operation of the ship, they were senior to, and in a sense responsible for the training and efficiency of the Active Officers.

The *ionomagnetic coils*, which served as the casings, "plates" and insulators of the gigantic condensers, were all located amidship on a center line, reaching clear through from the top to the bottom of the hull, and reaching from the forward to the rear rep-ray generators; that is, from points about 110 feet from bow and stern. The crew's quarters were arranged on both sides of the coils. To the outside of these, where the several decks touched the hull, were located the various pieces of *phone*, *scope* and *dis ray* apparatus.

The ship into which I was gazing with my *ultroscope* (at a telescopic and penetrative setting), carried a crew of perhaps 150 men all told. And except for the strained looks on their evil yellow faces I might have been tempted to believe I was looking on some Twenty-fifth Century pleasure excursion, for there was no running around nor appearance of activity.

CHAPTER VI

The Hans loved their ease, and despite the fact that this was a war ship, every machine and apparatus in it was equipped with a complement of seats and specially designed couches, in which officers and men reclined as they gazed at their viewplates, and manipulated the little sets of controls placed convenient to their hands.

* * * * *

The picture was a comic one to me, and I laughed, wondering how such soft creatures had held the sturdy and virile American race in complete subjection for centuries. But my laugh died as my mind grasped at the obvious explanation. These Hans were only soft physically. Mentally they were hard, efficient, ruthless, and conscienceless.

Impulsively I nosed my *swooper* down toward the ship and shot toward it at full rocket power. I had acted so swiftly that I had covered nearly half the distance toward the ship before my mind slowly drifted out of the daze of my emotion. This proved my undoing. Their scopeman saw me too quickly, for in heading directly at them I became easily visible, appearing as a steady, expanding point. Looking through their hull, I saw the crew of a *dis* ray generator come suddenly to attention. A second later their beam engulfed me.

For an instant my heart stood still. But the inertron shell of my swooper was impervious to the disintegrator ray. I was out of luck, however, so far as my control over my tiny ship was concerned. I had been hurtling in a direct line toward the ship when the beam found me. Now, when I tried to swerve out of the beam, the swooper responded but sluggishly to the shift I made in the rocket angle. I was, of course, traveling straight down a beam of vacuum. As my craft slowly nosed to the edge of the beam, the air rushing into this vacuum from all sides threw it back in again.

Had I shot my ship across one of these beams at right angles, my momentum would have carried me through with no difficulty. But I had no momentum now except in the line of the beam, and this being a vacuum now, my momentum, under full rocket power, was vastly increased. This realization gave me a second and more acute thrill.

CHAPTER VI

Would I be able to check my little craft in time, or would I, helpless as a bullet itself, crash through the shell of the Han ship to my own destruction?

I shut off my rocketmotor, but noticed no practical diminution of speed.

* * * * *

It was the fear of the Hans themselves that saved me. Through my ultroscope I saw sudden alarm on their faces, hesitation, a frantic officer in the control room jabbering into his phone. Then shakily the crew flipped their beam off to the side. The jar on my craft was terrific. Its nose caught the rushing tumble of air first, of course, and my tail sailing in a vacuum, swung around with a sickening wrench. My swooper might as well have been a barrel in the tumult of waters at the foot of Niagara. What was worse, the Hans kept me in that condition. Three of their beams were now playing in my direction, but not directly on me except for split seconds. Their technique was to play their beams around me more than on me, jerking them this way and that, so as to form vacuum pockets into which the air slapped and roared as the beams shifted, tossing me around like a chip.

Desperately I tried to bring my craft under control, to point its nose toward the Han ship and discharge an explosive rocket. Bitterly I cursed my self-confidence, and my impulsive action. An experienced pilot of the present age would have known better than to be caught shooting straight down a *dis* ray beam. He would have kept his ship shooting constantly at some angle to it, so that his momentum would carry him across it if he hit it. Too late I realized that there was more to the business of air fighting, than instinctive skill in guiding a swooper.

At last, when for a fraction of a second my nose pointed toward the Hans, I pressed the button of my rocket gun. I registered a hit, but not an accurate one. My projectile grazed an upper section of the ship's hull. At that it did terrific damage. The explosion battered in a section about fifty feet in diameter, partially destroying the top deck.

At the same instant I had shot my rocket, I had, in a desperate attempt to escape that turmoil of tumbling air, released a catch and dropped all

CHAPTER VI

that it was possible to drop of my ultron ballast. My swooper shot upward, like a bubble streaking for the surface of water.

I was free of the trap in which I had been caught, but unable to take advantage of the confusion which reigned on the Han ship.

I was as helpless to maneuver my ship now, in its up-rush, as when I had been tumbling in the air pockets. Moreover I was badly battered from plunging around in my shell like a pellet in a box, and partially unconscious.

I was miles in the air when I recovered myself. The swooper was steady enough now, but still rising, my instruments told me, and traveling in a general westward direction at full speed. Far below me was a sea of clouds, stretching from horizon to horizon, and through occasional breaks in its surface I could see still other seas of clouds at lower levels.

CHAPTER VII

Captured!

Certainly my situation was no less desperate. Unless I could find some method of compensating for my lost ballast, the inverse gravity of my inertron ship would hurl me continuously upward until I shot forth from the last air layer into space. I thought of jumping, and floating down on my inertron belt, but I was already too high for this. The air was too rarefied to permit breathing outside, though my little air compressors were automatically maintaining the proper density within the shell. If I could compress a sufficiently large quantity of air inside the craft, I would add to its weight. But there seemed little chance that I would myself be able to withstand sufficient compression.

I thought of releasing my inertron belt, but doubted whether this would be enough. Besides I might need the belt badly if I did find some method of bringing the little ship down, and it came too fast.

At last a plan came into my half-numbed brain that had some promise of success, though it was desperate enough. Cutting one of the hose pipes on my air compressor, and grasping it between my lips, I set to work to saw off the heads of the rivets that held the entire nose section of the swooper (inertron plates had to be grooved and riveted together, since the substance was impervious to heat and could not be welded). Desperately I sawed, hammered and chiseled, until at last with a wrench and a snap, the plate broke away.

The released nose of the ship shot upward. The rest began to drop with me. How fast I dropped I do not know, for my instruments went with the nose. Half fainting, I grimly clenched the rubber hose between my teeth, while the little compressor "carried on" nobly, despite the wrecked condition of the ship, giving me just enough air to keep my lungs from collapsing.

* * * * *

At last I shot through a cloud layer, and a long time afterward, it seemed, another. From the way in which they flashed up to meet me

CHAPTER VII

and to appear away above me, I must have been dropping like a stone.

At last I tried the rocket motor, very gently, to check my fall. The swooper was, of course, dropping tail first, and I had to be careful lest it turn over with a sharp blast from the motor, and dump me out.

Passing through the third layer of clouds I saw the earth beneath me. Then I jumped, pulling myself up through the jagged opening, and leaping upward while the remains of my ship shot away below me.

On approaching the ground I opened my chute-cape, to further check my fall, and landed lightly, with no further mishap. Whereupon I promptly threw myself down and slept, so exhausted was I with my experience.

It was not until the next morning that I awoke and gazed about me. I had come down in mountainous country. My intention was to get my bearing by tuning in headquarters with my ultrophone. But to my dismay I found the little battery disks had been torn from the earflaps of my helmet, though my chest-disk transmitter was still in place, and so far as I could see, in working order. I could report my experience, but could receive no reply.

I spent a half hour repeating my story and explanation on the headquarters channel, then once more surveyed my surroundings, trying to determine in which direction I had better leap. Then there came a stab of pain on the top of my head, and I dropped unconscious.

I regained consciousness to find myself, much to my surprise, a prisoner in the hands of a foot detachment of some thirty Hans. My surprise was a double one; first that they had not killed me instantly; second, that a detachment of them should be roaming this wild country afoot, obviously far from any of their cities, and with no ship hanging in the sky above them.

* * * * *

CHAPTER VII

As I sat up, their officer grunted with satisfaction and growled a guttural command. I was seized and pulled roughly to my feet by four soldiers, and hustled along with the party into a wooded ravine, through which we climbed sharply upward. I surmised, correctly as it turned out, that some projectile had grazed my head, and I was in such shape that if it had not been for the fact that my inertron belt bore most of my weight, they would have had to carry me. But as it was I made out well, and at the end of an hour's climb was beginning to feel like myself again, though the Han soldiers around me were puffing and drooping as men will, no matter how healthy, when they are totally unaccustomed to physical effort.

At length the party halted for a rest. I observed them curiously. Except for a few brief exciting moments at the time of our air raid on the intelligence office in Nu-Yok, I had seen no living specimens of this yellow race at close quarters.

They looked little like the Mongolians of the Twentieth Century, except for their slant eyes and round heads. The characteristic of the high cheek bones appeared to have been bred out of them, as were those of the relatively short legs and the muddy yellow skin. To call them yellow was more figurative than literal. Their skins were whiter than those of our own weather-tanned forest men. Nevertheless, their pigmentation was peculiar, and what there was of it looked more like a pale orange tint than the ruddiness of the Caucasian. They were well formed, but rather undersized and soft-looking, small-muscled and smooth-skinned, like young girls. Their features were finely chiseled, eyes beady, and nose slightly aquiline.

They were uniformed, not in close-fitting green or other shades of protective coloring, such as the unobtrusive gray of the Jersey Beaches or the leadened russet of the autumn uniforms of our people. Instead they wore loose fitting jackets of some silky material, and loose knee pants. This particular command had been equipped with form-moulded boots of some soft material that reached above the knee under their pants. They wore circular hats with small crowns and wide rims. Their loose jackets were belted at the waist, and they carried for weapons each man a knife, a short double-edged sword and what I took to be a form of magazine rocket gun. It was a rather bulky affair,

CHAPTER VII

short-barrelled, and with a pistol grip. It was obviously intended to be fired either from the waist position or from some sort of support, like the old machine guns. It looked, in fact, like a rather small edition of the Twentieth Century arm.

And have I mentioned the color of their uniforms? Their circular hats and pants were a bright yellow; their coats a flaming scarlet. What targets they were!

I must have chuckled audibly at the thought, for their commander who was seated on a folding stool one of his men had placed for him, glanced in my direction, and, at his arrogant gesture of command, I was prodded to my feet, and with my hands still bound, as they had been from the moment I recovered consciousness, I was dragged before him.

* * * * *

Then I knew what it was about these Hans that kept me in a turmoil of irritation. It was their sardonic, mocking, cruel smiles; smiles which left their stamp on their faces, even in repose. Now the commander was smiling tauntingly at me. When he spoke, it was in my own language.

"So!" he sneered. "You beasts have learned to laugh. You have gotten out of control in the last year or so. But that shall be remedied. In the meantime, a simple little surgical operation would make your smile a permanent one, reaching from ear to ear. But there, my orders are to deliver you and your equipment, all we have of it, intact. The Heaven-Born has had a whim."

"And who," I asked, "is this Heaven-Born?"

"San-Lan," he replied, "misbegotten spawn of the late High Priestess Nlui-Mok, and now Most Glorious Air Lord of All the Hans." He rolled out these titles with a bow of exaggerated respect toward the west, and in a tone of mockery. Those of his men who were near enough to hear, snickered and giggled.

CHAPTER VII

I was to learn that this amazing attitude of his was typical rather than exceptional. Strange as it may seem, no Han rendered any respect to another, nor expected it in return; that is, not genuine respect. Their discipline was rigid and cold-bloodedly heartless. The most elaborate courtesies were demanded and accorded among equals and from inferiors to superiors, but such was the intelligence and moral degradation of this remarkable race, that every one of them recognized these courtesies for what they were; they must of necessity have been hollow mockeries. They took pleasure in forcing one another to go through with them, each trying to outdo the other in cynical, sardonic thrusts, clothed in the most meticulously ceremonious courtesy. As a matter of fact, my captor, by this crude reference to the origin of his ruler, was merely proving himself a crude fellow, guilty of a vulgarity rather than of a treasonable or disrespectful remark. An officer of higher rank and better breeding, would have managed a clever innuendo, less direct, but equally plain.

I was about to ask him what part of the country we were in and where I was to be taken, when one of his men came running to him with a little portable electronophone, which he placed before him, with much bowing and scraping.

He conversed through this for a while, and then condescended to give me the information that a ship would soon be above us, and that I was to be transferred to it. In telling me this, he managed to convey, with crude attempts at mock-courtesy, that he and his men would feel relieved to be rid of me as a menace to health and sanitation, and would take exquisite joy in inflicting me upon the crew of the ship.

CHAPTER VIII

Hypnotic Torture

Some twenty minutes later the ship arrived. It settled down slowly into the ravine on its repeller rays until it was but a few feet above the tree tops. There it was stopped, and floated steadily, while a little cage was let down on a wire. Into this I was hustled and locked, whereupon the cage rose swiftly again to a hole in the bottom of the hull, into which it fitted snugly, and I stepped into the interior of a craft not unlike the one with which I had had my fateful encounter, the cage being unlocked.

The cabin in which I was confined was not an outside compartment, but was equipped with a number of viewplates.

The ship rose to a great height, and headed westward at such speed that the hum of the air past its smooth plates rose to a shrill, almost inaudible moan. After a lapse of some hours we came in sight of an impressive mountain range, which I correctly guessed to be the Rockies. Swerving slightly, we headed down toward one of the topmost pinnacles of the range, and there unfolded in one of the viewplates in my cabin a glorious view of Lo-Tan, the Magnificent, a fairy city of glistening glass spires and iridescent colors, piled up on sheer walls of brilliant blue, on the very tip of this peak.

Nor was there any sheen of shimmering disintegrator rays surrounding it, to interfere with the sparkling sight. So far-flung were the defenses of Lo-Tan, I found, that it was considered impossible for an American rocket gunner to get within effective range, and so numerous were the *dis* ray batteries on the mountain peaks and in the ravines, in this encircling line of defenses, drawn on a radius of no less than 100 miles, that even the largest craft, in the opinion of the Hans, could easily be brought to earth through air-pocketing tactics. And this, I was the more ready to believe after my own recent experience.

* * * * *

I spent two months as a prisoner in Lo-Tan. I can honestly say that during that entire time every attention was paid to my physical comfort.

CHAPTER VIII

Luxuries were showered upon me. But I was almost continuously subjected to some form of mental torture or moral assault. Most elaborately staged attempts at seduction were made upon me with drugs, with women. Hypnotism was resorted to. Viewplates were faked to picture to me the complete rout of American forces all over the continent. With incredible patience, and laboring under great handicaps, in view of the vigor of the American offensive, the Han intelligence department dug up the fact that somewhere in the forces surrounding Nu-Yok, I had left behind me Wilma, my bride of less than a year. In some manner, I will never tell how, they discovered some likeness of her, and faked an electronoscopic picture of her in the hands of torturers in Nu-Yok, in which she was shown holding out her arms piteously toward me, as though begging me to save her by surrender.

Surrender of what? Strangely enough, they never indicated that to me directly, and to this day I do not know precisely what they expected or hoped to get out of me. I surmise that it was information regarding the American sciences.

There was, however, something about the picture of Wilma in the hands of the torturers that did not seem real to me, and my mind still resisted. I remember gazing with staring eyes at that picture, the sweat pouring down my face, searching eagerly for some visible evidence of fraud and being unable to find it. It was the identical likeness of Wilma. Perhaps had my love for her been less great, I would have succumbed. But all the while I knew subconsciously that this was not Wilma. Product of the utmost of nobility in this modern virile, rugged American race, she would have died under even worse torture than these vicious Han scientists knew how to inflict, before she would have pleaded with me this way to betray my race and her honor.

But these were things that not even the most skilled of the Han hypnotists and psychoanalysts could drag from me. Their intelligence division also failed to pick up the fact that I was myself the product of the Twentieth Century and not the Twenty-fifth. Had they done so, it might have made a difference. I have no doubt that some of their most subtle mental assaults missed fire because of my own Twentieth Century "denseness." Their hypnotists inflicted many horrifying

CHAPTER VIII

nightmares on me, and made me do and say many things that I would not have done in my right senses. But even in the Twentieth Century we had learned that hypnotism cannot make a person violate his fundamental concepts of morality against his will, and steadfastly I steeled my will against them.

* * * * *

I have since thought that I was greatly aided by my newness to this age. I have never, as a matter of fact, become entirely attuned to it. And even today I confess to a longing wish that man might travel backward as well as forward in time. Now that my Wilma has been at rest these many years, I wish that I might go back to the year 1927, and take up my old life where I left it off, in the abandoned mine near Scranton.

And at the period of which I speak, I was less attuned than now to the modern world. Real as my life was, and my love for my wife, there was much about it all that was like a dream, and in the midst of my tortures by the Hans, this complex--this habit of many months--helped me to tell myself that this, too, was all a dream, that I must not succumb, for I would wake up in a moment.

And so they failed.

More than that, I think I won something nearer to genuine respect from those around me than any other Hans of that generation accorded to anybody.

Among these was San-Lan himself, the ruler. In the end it was he who ordered the cessation of these tortures, and quite frankly admitted to me his conviction that they had been futile and that I was in many senses a super-man. Instead of having me executed, he continued to shower luxuries and attentions on me, and frequently commanded my attendance upon him.

Another was his favorite concubine, Ngo-Lan, a creature of the most alluring beauty; young, graceful and most delicately seductive, whose skill in the arts and sciences put many of their doctors to shame. This

CHAPTER VIII

creature, his most prized possession, San-Lan with the utmost moral callousness ordered to seduce me, urging her to apply without stint and to its fullest extent, her knowledge of evil arts. Had I not seen the naked horror of her soul, that she let creep into her eyes for just one unguarded instant, and had it not been for my conviction of Wilma's faith in me, I do not know what--but suffice it to say that I resisted this assault also.

Had San-Lan only known it, he might have had a better chance of breaking down my resistance through another bit of femininity in his household, the little nine-year-old Princess Lu-Yan, his daughter.

* * * * *

I think San-Lan held something of real affection for this sprightly little mite, who in spite of the sickening knowledge of rottenness she had already acquired at this early age, was the nearest thing to innocence I found in Lo-Tan. But he did not realize this, and could not; for even the most natural and fundamental affection of the human race, that of parents for their offspring, had been so degraded and suppressed in this vicious Han civilization as to be unrecognizable. Naturally San-Lan could not understand the nature of my pity for this poor child, nor the fact that it might have proved a weak spot in my armor. But had he done so, I truly believe he would have been ready to inflict degradation, torture and even death upon her, to make me surrender the information he wanted.

Yet this man, perverted product of a morally degraded race, had about him something of true dignity; something of sincerity, in a warped, twisted way. There were times when he seemed to sense vaguely, gropingly, wonderingly, that he might have a soul.

The Han philosophy for centuries had not admitted the existence of souls. Its conception embraced nothing but electrons, protons and molecules, and still was struggling desperately for some shred of evidence that thoughts, will power and consciousness of self were nothing but chemical reactions. However, it had gotten no further than the negative knowledge we had in the Twentieth Century, that a sick body dulls consciousness of the material world, and that knowledge,

CHAPTER VIII   44

which all mankind has had from the beginning of time, that a dead body means a departed consciousness. They had succeeded in producing, by synthesis, what appeared to be living tissues, and even animals of moderately complex structure and rudimentary brains, but they could not give these creatures the full complement of life's characteristics, nor raise the brains to more than mechanical control of muscular tissues.

It was my own opinion that they never could succeed in doing so. This opinion impressed San-Lan greatly. I had expected him to snort his disgust, as the extreme school of evolutionists would have done in the Twentieth Century. But the idea was as new to him and the scientists of his court as Darwinism was to the late Nineteenth and early Twentieth Centuries. So it was received with much respect. Painfully and with enforced mental readjustments, they began a philosophical search for excuses and justifications for the idea.

* * * * *

All of this amused me greatly, for of course neither the newness nor the orthodoxy of a hypothesis will make it true if it is not true, nor untrue if it is true. Nor could the luck or will-power, with which I had resisted their hypnotists and psychoanalysts, make what might or might not be a universal fact one whit more or less of a fact than it really was. But the prestige I had gained among them, and the novelty of my expressed opinion carried much weight with them.

Yet, did not even brilliant scientists frequently exhibit the same lack of logic back in the Twentieth Century? Did not the historians, the philosophers of ancient Greece and Rome show themselves to be the same shrewd observers as those of succeeding centuries, the same masters of the logical and slaves of the illogical?

After all, I reflected, man makes little progress within himself. Through succeeding generations he piles up those resources which he possesses outside of himself, the tools of his hands, and the warehouses of knowledge for his brain, whether they be parchment manuscripts, printed book, or electronorecordographs. For the rest he is born today, as in ancient Greece, with a blank brain, and struggles

CHAPTER VIII

through to his grave, with a more or less beclouded understanding, and with distinct limitations to what we used to call his "think tank."

* * * * *

This particular reflection of mine proved unpopular with them, for it stabbed their vanity, and neither my prestige nor the novelty of the idea was sufficient salve. These Hans for centuries had believed and taught their children that they were a super-race, a race of destiny. Destined to Whom, for What, was not so clear to them; but nevertheless destined to "elevate" humanity to some sort of super-plane. Yet through these same centuries they had been busily engaged in the extermination of "weaklings," whom, by their very persecutions, they had turned into "super men," now rising in mighty wrath to destroy them; and in reducing themselves to the depths of softening vice and flabby moral fiber. Is it strange that they looked at me in amazed wonder when I laughed outright in the midst of some of their most serious speculations?

CHAPTER IX

The Fall of Nu-Yok

My position among the Hans, in this period, was a peculiar one. I was at once a closely guarded prisoner and an honored guest. San-Lan told me frankly that I would remain the latter only so long as I remained an object of serious study or mental diversion to himself or his court. I made bold to ask him what would be done with me when I ceased to be such.

"Naturally," he said, "you will be eliminated. What else? It takes the services of fifteen men altogether, to guard you; and men, you understand, cannot be produced and developed in less than eighteen years." He meditated frowningly for a moment. "That, by the way, is something I must take up with the Birth and Educational Bureau. They must develop some method of speeding growth, even at the cost of mental development. With your wild forest men getting out of hand this way, we are going to need greater resources of population, and need them badly.

"But," he continued more lightly, "there seems to be no need for you to disturb yourself over the prospect at present. It is true you have been able to resist our psychoanalysts and hypnotists, and so have no value to us from the viewpoint of military information, but as a philosopher, you have proved interesting indeed."

He broke off to give his attention to a gorgeously uniformed official who suddenly appeared on the large viewplate that formed one wall of the apartment. So perfectly did this mechanism operate, that the man might have been in the room with us. He made a low obeisance, then rose to his full height and looked at his ruler with malicious amusement.

"Heaven-Born," he said, "I have the exquisite pain of reporting bad news."

San-Lan gave him a scathing look. "It will be less unpleasant if I am not distracted by the sight of you while you report."

CHAPTER IX

At this the man disappeared, and the viewplate once more presented its normal picture of the mountains north of Lo-Tan; but the voice continued:

"Heaven-Born, the Nu-Yok fleet has been destroyed, the city is in ruins, and the newly formed ground brigades, reduced to 10,000 men, have taken refuge in the hills of Ron-Dak (the Adirondacks) where they are being pressed hard by the tribesmen, who have surrounded them."

* * * * *

For an instant San-Lan sat as though paralyzed. Then he leaped to his feet, facing the viewplate.

"Let me see you!" he snarled. Instantly the mountain view disappeared and the Intelligence Officer appeared again, this time looking a little frightened.

"Where is Lui-Lok?" he shouted. "Cut him in on my north plate. The commander who loses his city dies by torture. Cut him in. Cut him in!"

"Heaven-Born, Lui-Lok committed suicide. He leaped into a ray, when the rockets of the tribesmen began to penetrate the ray-wall. Lip-Hung is in command of the survivors. We have just had a message from him. We could not understand all of it. Reception was very weak because he is operating with emergency apparatus on Bah-Flo power. The Nu-Yok power broadcast plant has been blown up. Lip-Hung begs for a rescue fleet."

San-Lan, his expression momentarily becoming more vicious, now was striding up and down the room, while the poor wretch in the viewplate, thoroughly scared at last, stood trembling.

"What!" shrieked the tyrant. "He begs a rescue. A rescue of what? Of 10,000 beaten men and nothing better than makeshift apparatus? No fleet? No city? I give him and his 10,000 to the tribesmen! They are of no use to us now! Get out! Vanish! No, wait! Have any of the beasts' rockets penetrated the ray-walls of other cities?"

CHAPTER IX

"No, Heaven-Born, no. It is only at Nu-Yok that the tribesmen used rockets sheathed in the same mysterious substance they use on their little aircraft and which cannot be disintegrated by the ray." (He meant inertron, of course.)

San-Lan waved his hand in dismissal. The officer dissolved from view, and the mountains once more appeared, as though the whole side of the room were of glass.

More slowly he paced back and forth. He was the caged tiger now, his face seamed with hate and the desperation of foreshadowed doom.

"Driven out into the hills," he muttered to himself. "Not more than 10,000 of them left. Hunted like beasts--and by the very beasts we ourselves have hunted for centuries. Cursed be our ancestors for letting a single one of the spawn live!" He shook his clenched hands above his head. Then, suddenly remembering me, he turned and glared.

"Forest man, what have you to say?" he demanded.

Thus confronted, there stole over me that same detached feeling that possessed me the day I had been made Boss of the Wyomings.

"It is the end of the Air Lords of Han," I said quietly. "For five centuries command of the air has meant victory. But this is so no longer. For more than three centuries your great, gleaming cities have been impregnable in all their arrogant visibility. But that day is done also. Victory returns once more to the ground, to men invisible in the vast expanse of the forest which covers the ruins of the civilization destroyed by your ancestors. Ye have sown destruction. Ye shall reap it!

"Your ancestors thought they had made mere beasts of the American race. Physically you did reduce them to the state of beasts. But men do have souls, San-Lan, and in their souls the Americans still cherished the spark of manhood, of honor, of independence. While the Hans have degenerated into a race of sleek, pampered beasts themselves, they have unwittingly bred a race of super-men out of

CHAPTER IX

those they sought to make animals. You have bred your own destruction. Your cities shall be blasted from their foundations. Your air fleets shall be brought crashing to earth. You have your choice of dying in the wreckage, or of fleeing to the forests, there to be hunted down and killed as you have sought to destroy us!"

And the ruler of all the Hans shrank back from my outstretched finger as though it had been in truth the finger of doom.

But only for a moment. Suddenly he snarled and crouched as though to spring at me with his bare hands. By a mighty convulsion of the will he regained control of himself, however, and assumed a manner of quiet dignity. He even smiled--a slow, crooked smile.

"No," he said, answering his own thought. "I will not have you killed now. You shall live on, my honored guest, to see with your own eyes how we shall exterminate your animal-brethren in their forests. With your own ears you shall hear their dying shrieks. The cold science of Han is superior to your spurious knowledge. We have been careless. To our cost we have let you develop brains of a sort. But we are still superior. We shall go down into the forests and meet you. We shall beat you in your own element. When you have seen and heard this happen, my Council shall devise for you a death by scientific torture, such as no man in the history of the world has been honored with."

* * * * *

I must digress here a bit from my own personal adventures to explain briefly how the fall of Nu-Yok came about, as I learned it afterward.

Upon my capture by the Hans, my wife, Wilma, courageously had assumed command of my Gang, the Wyomings.

Boss Handan, of the Winslows, who was directing the American forces investing Nu-Yok, contented himself for several weeks with maintaining our lines, while waiting for the completion of the first supply of inertron-jacketed rockets. At last they arrived with a limited quantity of very high-powered atomic shells, a trifle over a hundred of them to be exact. But this number, it was estimated, would be enough

CHAPTER IX

to reduce the city to ruins. The rockets were distributed, and the day for the final bombardment was set.

The Hans, however, upset Handan's plans by launching a ground expedition up the west bank of the Hudson. Under cover of an air raid to the southwest, in which the bulk of their ships took part, this ground expedition shot northward in low-flying ships.

The raiding air fleet ploughed deep into our lines in their famous "cloud-bank" formation, with down-playing disintegrator rays so concentrated as to form a virtual curtain of destruction. It seared a scar path a mile and a half wide fifteen miles into our territory.

Everyone of our rocket gunners caught in this section was annihilated. Altogether we lost several hundred men and girls.

Gunners to each side of the raiding ships kept up a continuous fire on them. Most of the rockets were disintegrated, for Handan would not permit the use of the inertron rockets against the ships. But now and then one found its way through the playing beams, hit a repeller ray and was hurled up against a Han ship, bringing it crashing down.

The orders that Handan barked into his ultrophone were, of course, heard by every long-gunner in the ring of American forces around the city, and nearly all of them turned their fire on the Han airfleet, with the exception of those equipped with the inertron rockets.

These latter held to the original target and promptly cut loose on the city with a shower of destruction which the disintegrator-ray walls could not stop. The results produced awe even in our own ranks.

* * * * *

Where an instant before had stood the high-flung masses and towers of Nu-Yok, gleaming red, blue and gold in the brilliant sunlight, and shimmering through the iridescence of the ray "wall," there was a seething turmoil of gigantic explosions.

CHAPTER IX

Surging billows of debris were hurled skyward on gigantic pulsations of blinding light, to the detonations that shook men from their feet in many sections of the American line seven and eight miles away.

As I have said, there were only some hundred of the inertron rockets among the Americans, long and slender, to fit the ordinary guns, but the atomic laboratories hidden beneath the forests, had outdone themselves in their construction. Their release of atomic force was nearly 100 per cent, and each one of them was equal to many hundred tons of trinitrotoluol, which I had known in the First World War, five hundred years before, as "T.N.T."

It was all over in a few seconds. Nu-Yok had ceased to exist, and the waters of the bay and the rivers were pouring into the vast hole where a moment before had been the rocky strata beneath lower Manhattan.

Naturally, with the destruction of the city's power-broadcasting plant the Han air fleet had plunged to earth.

But the ships of the ground expedition up the river, hugging the tree tops closely, had run the gauntlet of the American long-gunners who were busily shooting at the other Han fleet, high in the air to the southwest, and about half of them had landed before their ships were robbed of their power. The other half crashed, taking some 10,000 or 12,000 Han troops to destruction with them. But from those which had landed safely, emerged the 10,000 who now were the sole survivors of the city, and who took refuge in wooded fastnesses of the Adirondacks.

* * * * *

The Americans with their immensely greater mobility, due to their jumping belts and their familiarity with the forest, had them ringed in within twenty-four hours.

But owing to the speed of the maneuvers, the lines were not as tightly drawn as they might have been, and there was considerable scattering of both American and Han units. The Hans could make only the weakest short-range use of their newly developed disintegrator-ray

CHAPTER IX

field units, since they had only distant sources of power-broadcast on which to draw. On the other hand, the Americans could use their explosive rockets only sparingly for fear of hitting one another.

So the battle was finished in a series of desperate hand-to-hand encounters in the ravines and mountain slopes of the district.

The Mifflins and Altoonas, themselves from rocky, mountainous sections, gave a splendid account of themselves in this fighting, leaping to the craggy slopes above the Hans, and driving them down into the ravines, where they could safely concentrate on them the fire of depressed rocket guns.

The Susquannas, with their great inertron shields, which served them well against the weak rays of the Hans, pressed forward irresistibly every time they made a contact with a Han unit, their short-range rocket guns sending a hail of explosive destruction before them.

But the Delawares, with their smaller shields, inertron leg-guards and helmets, and their ax-guns, made faster work of it. They would rush the Hans, shooting from their shields as they closed in, and finish the business with their ax-blades and the small rocket guns that formed the handles of their axes.

It was my own unit of Wyomings, equipped with bayonet guns not unlike the rifles of the First World War, that took the most terrible toll from the Hans.

They advanced at the double, laying a continuous barrage before them as they ran, closing with the enemy in great leaps, cutting, thrusting and slicing with those terrible double-ended weapons in a vicious efficiency against which the Hans with their swords, knives and spears were utterly helpless.

And so my prediction that the war would develop hand-to-hand fighting was verified at the outset.

None of the details of this battle of the Ron-Daks were ever known in Lo-Tan. Not more than the barest outlines of the destruction of the survivors of Nu-Yok were ever received by San-Lan and his Council. And of course, at that time I knew no more about it than they did.

CHAPTER X

Life in Lo-Tan, the Magnificent

San-Lan's attitude toward me underwent a change. He did not seek my company as he had done before, and so those long discussions and mental duels in which we pitted our philosophies against each other came to an end. I was, I suspected, an unpleasant reminder to him of things he would rather forget, and my presence was an omen of impending doom. That he did not order my execution forthwith was due, I believe to a sort of fascination in me, as the personification of this (to him) strange and mysterious race of super-men who had so magically developed overnight from "beasts" of the forest.

But though I saw little of him after this, I remained a member of his household, if one may speak of a "household" where there is no semblance of house.

The imperial apartments were located at the very summit of the Imperial Tower, the topmost pinnacle of the city, itself clinging to the sides and peak of the highest mountain in that section of the Rockies. There were days when the city seemed to be built on a rugged island in the midst of a sea of fleecy whiteness, for frequently the cloud level was below the peak. And on such days the only visual communications with the world below was through the viewplates which formed nearly all the interior walls of the thousands of apartments (for the city was, in fact, one vast building) and upon which the tenants could tune in almost any views they wished from an elaborate system of public television and projectoscope broadcasts.

Every Han city had many public-view broadcasting stations, operating on tuning ranges which did not interfere with other communication systems. For slight additional fees a citizen in Lo-Tan might, if he felt so inclined, "visit" the seashore, or the lakes or the forests of any part of the country, for when such scene was thrown on the walls of an apartment, the effect was precisely the same as if one were gazing through a vast window at the scene itself.

CHAPTER X

It was possible too, for a slightly higher fee, to make a mutual connection between apartments in the same or different cities, so that a family in Lo-Tan, for instance, might "visit" friends in Fis-Ko (San Francisco) taking their apartment, so to speak, along with them; being to all intents and purposes separated from their "hosts" only by a big glass wall which interfered neither with vision nor conversation.

These public view and visitation projectoscopes explain that utter depth of laziness into which the Hans had been dragged by their civilization. There was no incentive for anyone to leave his apartment unless he was in the military or air service, or a member of one of the repair services which from time to time had to scoot through the corridors and shafts of the city, somewhat like the ancient fire departments, to make some emergency repair to the machinery of the city or its electrical devices.

* * * * *

Why should he leave his house? Food, wonderful synthetic concoctions of any desired flavor and consistency (and for additional fee conforming to the individual's dietary prescription) came to him through a shaft, from which his tray slid automatically on to a convenient shelf or table.

At will he could tune in a theatrical performance of talking pictures. He could visit and talk with his friends. He breathed the freshest of filtered air right in his own apartment, at any temperature he desired, fragrant with the scent of flowers, the aromatic smell of the pine forests or the salt tang of the sea, as he might prefer. He could "visit" his friends at will, and though his apartment actually might be buried many thousand feet from the outside wall of the city, it was none the less an "outside" one, by virtue of its viewplate walls. There was even a tube system, with trunk, branch and local lines and an automagnetic switching system, by which articles within certain size limits could be despatched from any apartment to any other one in the city.

The women actually moved about through the city more than the men, for they had no fixed duties. No work was required of them, and though nominally free, their dependence upon the government pension for their necessities and on their "husbands" (of the moment) for their luxuries, reduced them virtually to the condition of slaves.

Each had her own apartment in the Lower City, with but a single small viewplate, very limited "visitation" facilities, and a minimum credit for food and clothing. This apartment was assigned to her on graduation from the State School, in which she had been placed as an infant, and it remained hers so long as she lived, regardless of whether she occupied it or not. At the conclusion of her various "marriages" she would return there, pending her endeavors to make a new match. Naturally, as her years increased, her returns became more frequent and her stay of longer duration, until finally, abandoning hope of making another match, she finished out her days there, usually in drunkenness and whatever other forms of cheap dissipation she could afford on her dole, starving herself.

Men also received the same State pension, sufficient for the necessities but not for the luxuries of life. They got it only as an old-age pension, and on application.

* * * * *

When boys graduated from the State School they generally were "adopted" by their fathers and taken into the latter's households, where they enjoyed luxuries far in excess of their own earning power. It was not that their fathers wasted any affection on them, for as I have explained before, the Hans were so morally atrophied and scientifically developed that love and affection, as we Americans knew them, were unexperienced or suppressed emotions with them. They were replaced by lust and pride of possession. So long as it pleased a father's vanity, and he did not miss the cost, he would keep a son with him, but no longer.

Young men, of course, started to work at the minimum wage, which was somewhat higher than the pension. There was work for everybody in positions of minor responsibility, but very little hard work.

Upon receiving his appointment from one or another of the big corporations which handled the production and distribution of the vast

CHAPTER X

community (the shares of which were pooled and held by the government--that is, by San-Lan himself--in trust for all the workers, according to their positions) he would be assigned to an apartment-office, or an apartment adjoining the group of offices in which he was to have his desk. Most of the work was done in single apartment-offices.

The young man, for instance, might recline at his ease in his apartment near the top of the city, and for three or four hours a day inspect, through his viewplate and certain specially installed apparatus, the output of a certain process in one of the vast automatically controlled food factories buried far underground beneath the base of the mountain, where the moan of its whirring and throbbing machinery would not disturb the peace and quiet of the citizens on the mountain top. Or he might be required simply to watch the operation of an account machine in an automatic store.

There is no denying that the economic system of the Hans was marvelous. A suit of clothes, for instance, might be delivered in a man's apartment without a human hand having ever touched it.

Having decided that he wished a suit of a given general style, he would simply tune in a visual broadcast of the display of various selections, and when he had made his choice, dial the number of the item and press the order button. Simultaneously the charge would be automatically made against his account number, and credited as a sale on the automatic records of that particular factory in the account house. And his account plate, hidden behind a little wall door, would register his new credit balance. An automatically packaged suit that had been made to style and size-standard by automatic machinery from synthetically produced material, would slip into the delivery chute, magnetically addressed, and in anywhere from a few seconds to thirty minutes or so, according to the volume of business in the chutes, and drop into the delivery basket in his room.

* * * * *

Daily his wages were credited to his account, and monthly his share of the dividends likewise (according to his position) from the Imperial

CHAPTER X

Investment Trust, after deduction of taxes (through the automatic bookkeeping machines) for the support of the city's pensioners and whatever sum San-Lan himself had chosen to deduct for personal expenses and gratuities.

A man could not bequeath his ownership interest in industry to his son, for that interest ceased with his death, but his credit accumulation, on which interest was paid, was credited to his eldest recorded son as a matter of law.

Since many of these credit fortunes (The Hans had abandoned gold as a financial basis centuries before) were so big that they drew interest in excess of the utmost luxury costs of a single individual, there was a class of idle rich consisting of eldest sons, passing on these credit fortunes from generation to generation. But younger sons and women had no share in these fortunes, except by the whims and favor of the "Man-Dins" (Mandarins), as these inheritors were known.

These Man-Dins formed a distinct class of the population, and numbered about five percent of it. It was distinct from the Ku-Li (coolie) or common people, and from the "Ki-Ling" or aristocracy composed of those more energetic men (at least mentally more energetic) who were the active or retired executive heads of the various industrial, educational, military or political administrations.

A man might, if he so chose, transfer part of his credit to a woman favorite, which then remained hers for life or until she used it up, and of course, the prime object of most women, whether as wives, or favorites, was to beguile a settlement of this sort out of some wealthy man.

When successful in this, and upon reassuming her freedom, a woman ranked socially and economically with the Man-Dins. But on her death, whatever remained of her credit was transferred to the Imperial fund.

When one considers that the Hans, from the days of their exodus from Mongolia and their conquest of America, had never held any ideal of monogamy, and the fact that marriage was but a temporary formality which could be terminated on official notice by either party, and that after all it gave a woman no real rights or prerogatives that could not be terminated at the whim of her husband, and established her as nothing but the favorite of his harem, if he had an income large enough to keep one, or the most definitely acknowledged of his favorites if he hadn't, it is easy to see that no such thing as a real family life existed among them.

Free women roamed the corridors of the city, pathetically importuning marriage, and wives spent most of the time they were not under their husbands' watchful eyes in flirtatious attempts to provide themselves with better prospects for their next marriages.

* * * * *

Naturally the biggest problem of the community was that of stimulating the birth rate. The system of special credits to mothers had begun centuries before, but had not been very efficacious until women had been deprived of all other earning power, and even at the time of which I write it was only partially successful, in spite of the heavy bounties for children. It was difficult to make the bounties sufficiently attractive to lure the women from their more remunerative light flirtations. Eugenic standards also were a handicap.

As a matter of fact, San-Lan had under consideration a revolutionary change in economic and moral standards, when the revolt of the forest men upset his delicately laid plans, for, as he had explained to me, it was no easy thing to upset the customs of centuries in what he was pleased to call the "morals" of his race.

He had another reason too. The physically active men of the community were beginning to acquire a rather dangerous domination. These included men in the army, in the airships, and in those relatively few civilian activities in which machines could not do the routine work and thinking. Already common soldiers and air crews demanded and received higher remuneration than all except the highest of the Ki-Ling, the industrial and scientific leaders, while mechanics and repairmen who could, and would, work hard physically, commanded higher incomes than Princes of the Blood, and though constituting only a fraction of one per cent of the population they actually dominated the city. San-Lan dared take no important step in the development of the industrial and military system without consulting their council or Yun-Yun (Union), as it was known.

Socially the Han cities were in a chaotic condition at this time, between morals that were not morals, families that were not families, marriages that were not marriages, children who knew no homes, work that was not work, eugenics that didn't work; Ku-Lis who envied the richer classes but were too lazy to reach out for the rewards freely offered for individual initiative; the intellectually active and physically lazy Ki-Lings who despised their lethargy; the Man-Din drones who regarded both classes with supercilious toleration; the Princes of the Blood, arrogant in their assumption of a heritage from a Heaven in which they did not believe; and finally the three castes of the army, air and industrial repair services, equally arrogant and with more reason in their consciousness of physical power.

* * * * *

The army exercised a cruelly careless and impartial police power over all classes, including the airmen, when the latter were in port. But it did not dare to touch the repair men, who, so far as I could ever make out, roamed the corridors of the city at will during their hours off duty, wreaking their wills on whomever they met, without let or hindrance.

Even a Prince of the Blood would withdraw into a side corridor with his escort of a score of men, to let one of these labor "kings" pass, rather than risk an altercation which might result in trouble for the government with the Yun-Yun, regardless of the rights and wrongs of the case, unless a heavy credit transference was made from the balance of the Prince to that of the worker. For the machinery of the city could not continue in operation a fortnight, before some accident requiring delicate repair work would put it partially out of commission. And the Yun-Yun was quick to resent anything it could construe as a slight on one of its members.

In the last analysis it was these Yun-Yun men, numerically the smallest of the classes, who ruled the Han civilization, because for all practical purposes they controlled the machinery on which that civilization depended for its existence.

Politically, San-Lan could balance the organizations of the army and the air fleets against each other, but he could not break the grip of the repairmen on the machinery of the cities and the power broadcast plants.

CHAPTER XI

The Forest Men Attack

Many times during the months I remained prisoner among the Hans I had tried to develop a plan of escape, but could conceive of nothing which seemed to have any reasonable chance of success.

While I was allowed almost complete freedom within the confines of the city, and sometimes was permitted to visit even the military outposts and disintegrator ray batteries in the surrounding mountains, I was never without a guard of at least five men under the command of an officer. These men were picked soldiers, and they were armed with powerful though short-range disintegrator-ray pistols, capable of annihilating anything within a hundred feet. Their vigilance never relaxed. The officer on duty kept constantly at my side, or a couple of paces behind me, while certain of the others were under strict orders never to approach within my reach, nor to get more than forty feet away from me. The thought occurred to me once to seize the officer at my side and use him as a shield, until I found that the guard were under orders to destroy both of us in such a case.

So in this fashion I roamed the city corridors, wherever I wished. I visited the great factories at the bottom of the shafts that led to the base of the mountain, where, unattended by any mechanics, great turbines whirred and moaned, giant pistons plunged back and forth, and immense systems of chemical vats, piping and converters, automatically performed their functions with the assistance of no human hand, but under the minute television inspection of many perfumed dandies reclining at their ease before viewplates in their apartment offices in the city, that clung to the mountain peak far above.

There were just two restrictions on my freedom of movement. I was allowed nowhere near the power-broadcasting station on the peak, nor the complement of it which was buried three miles below the base of the mountain. And I was never allowed to approach within a hundred feet of any disintegrator ray machine when I visited the military outposts in the surrounding mountains.

CHAPTER XI

I first noticed the "escape tunnels" one day when I had descended to the lowest level of all, the location of the Electronic Plant, where machines, known as "reverse disintegrators," fed with earth and crushed rock by automatic conveyors, subjected this material to the disintegrator ray, held the released electrons captive within their magnetic fields and slowly refashioned them into supplies of metals and other desired elements.

My attention was attracted to the tunnels by the unusual fact that men were busily entering and leaving them. Almost the entire repair force seemed to be concentrated here. Stocky, muscular men they were, with the same modified Oriental countenances as the rest of the Hans, but with a certain ruggedness about them that was lacking in the rest of the indolent population. They sweated as they labored over the construction of magnetic cars evidently designed to travel down these tunnels, automatically laying pipe lines for ventilation and temperature control. The tunnels themselves appeared to have been driven with disintegrator rays, which could bore rapidly through the solid rock, forming glassy iridescent walls as they bored, and involving no problem of debris removal.

* * * * *

I asked San-Lan about it the next time I saw him, for the officer of my guard would give me no information.

The supreme ruler of the Hans smiled mockingly.

"There is no reason why you should not know their purpose," he said, "for you will never be able to stop our use of them. These tunnels constitute the road to a new Han era. Your forest men have turned our cities into traps, but they have not trapped our minds and our powers over Nature. We are masters still; masters of the world, and of the forest men.

"You have revolutionized the tactics of warfare with your explosive rockets and your strategy of fighting from concealed positions, miles away, where we cannot find you with our beams. You have driven our ships from the air, and you may destroy our cities. But we shall be

CHAPTER XI

gone.

"Down these tunnels we shall depart to our new cities, deep under ground, and scattered far and wide through the mountains. They are nearly completed now.

"You will never blast us out of these, even with your most powerful explosives, because they will be more difficult for you to find than it is for us to locate a forest gunner somewhere beneath his leafy screen of miles of trees, and because they will be too far underground."

"But," I objected, "man cannot live and flourish like a mole continually removed from the light of day, without the health-giving rays of the sun, which man needs."

"No?" San-Lan jeered. "Wild tribesmen might not be able to, but we are a civilization. We shall make our own sunlight to order in the bowels of the earth. If necessary, we can manufacture our air synthetically; not the germ-laden air of Nature, but absolutely pure air. Our underground cities will be heated or refrigerated artificially as conditions may require. Why should we not live underground if we desire? We produce all our needs synthetically.

"Nor will you be able to locate our cities with electronic indicators.

"You see, Rogers, I know what is in your mind. Our scientists have planned carefully. All our machinery and processes will be shielded so that no electronic disturbances will exist at the surface.

"And then, from our underground cities we will emerge at leisure to wage merciless war on your wild men of the forest, until we have at last done what our forefathers should have done, exterminated them to the last beast."

* * * * *

He thrust his jeering face close to mine. "Have you any answer to that?" he demanded.

CHAPTER XI

My impulse was to plant my fist in his face, for I could think of no other answer. But I controlled myself, and even forced a hearty laugh, to irritate him.

"It is a fine plan," I admitted, "but you will not have time to carry it through. Long before you can complete your new cities you will have been destroyed."

"They will be completed within the week," he replied triumphantly. "We have not been asleep, and our mechanical and scientific resources make us masters of time as well as the earth. You shall see."

Naturally I was worried. I would have given much if I could have passed this information on to our chiefs.

But two days later a mighty exultation arose within me, when from far to the east and also to the south there came the rolling and continuous thunder of rocket fire. I was in my own apartment at the time. The Han captain of my guard was with me, as usual, and two guards stood just within the door. The others were in the corridor outside. And as soon as I heard it, I questioned my jailer with a look. He nodded assent, and I did what probably every disengaged person in Lo-Tan did at the same moment, tuned in on the local broadcast of the Military Headquarters View and Control Room.

It was as though the side wall of my apartment had dissolved, and we looked into a large room or office which had no walls or ceiling, these being replaced by the interior surface of a hemisphere, which was in fact a vast viewplate on which those in the room could see in every direction. Some 200 staff officers had their desks in this room. Each desk was equipped with a system of small viewplates of its own, and each officer was responsible for a given directional section of the "map," and busied himself with teleprojectoscope examination of it, quite independently of the general view thrown on the dome plate.

At a raised circular desk in the center, which was composed entirely of viewplates, sat the Executive Marshal, scanning the hemisphere, calling occasionally for telescopic views of one section or another on his desk plates, and noting the little pale green signal lights that

CHAPTER XI

flashed up as Sector Observers called for his attention.

* * * * *

Members of Strategy Board, Base Commanders of military units, and San-Lan himself, I understood, sat at similar desks in their private offices, on which all these views were duplicated, and in constant verbal and visual communication with one another and with the Executive Marshal.

The particular view which appeared on my own wall fortunately showed the east side of the dome viewplate and in one corner of my picture appeared the Executive Marshal himself.

Although I was getting a viewplate picture of a viewplate picture, I could see the broad, rugged valley to the east plainly, and the relatively low ridge beyond, which must have been some thirty miles away.

It was beyond this, evidently far beyond it, that the scene of the action was located, for nothing showed on the plate but a misty haze permeated by indefinite and continuous pulsations of light, and against which the low mountain ridge stood out in bold relief.

Somewhere on the floor of the Observation room, of course, was a Sector Observer who was looking beyond that ridge, probably through a projectoscope station in the second or third "circle," located perhaps on that ridge or beyond it.

At the very moment I was wishing for his facilities the Executive Marshal leaned over to a microphone and gave an order in a low tone. The hemispherical view dissolved, and another took its place, from the third circle. And the view was now that which would be seen by a man standing on the low distant ridge.

There was another broad valley, a wide and deep canyon, in fact, and beyond this still another ridge, the outlines of which were already beginning to fade into the on-creeping haze of the barrage. The flashes of the great detonating rockets were momentarily becoming more vivid.

"That's the Gok-Man ridge," mused the Han officer beside me in the apartment, "and the Forest Men must be more than fifty miles beyond that."

"How do you figure that?" I asked curiously.

"Because obviously they have not penetrated our scout lines. See that line of observers nearest the dome itself. They're all busy with their desk plates. They're in communication with the scout line. The scout line broadcast is still in operation. It looks as though the line is still unpierced, but the tribesmen's rockets are sailing over and falling this side of it."

All through the night the barrage continued. At times it seemed to creep closer and then recede again. Finally it withdrew, pulling back to the American lines, to alternately advance and recede. At last I went to sleep. The Han officer seemed to be a relatively good-natured fellow, for one of his race, and he promised to awake me if anything further of interest took place.

He didn't though. When I awoke in the morning, he gave me a brief outline of what had happened.

It was pieced together from his own observations and the public news broadcast.

CHAPTER XII

The Mysterious "Air Balls"

The American barrage had been a long distance bombardment, designed, apparently, to draw the Han disintegrator ray batteries into operation and so reveal their positions on the mountain tops and slopes, for the Hans, after the destruction of Nu-Yok, had learned quickly that concealment of their positions was a better protection than a surrounding wall of disintegrator rays shooting up into the sky.

The Hans, however, had failed to reply with disintegrator rays. For already this arm, which formerly they had believed invincible, was being restricted to a limited number of their military units, and their factories were busy turning out explosive rockets not dissimilar to those of the Americans in their motive power and atomic detonation. They had replied with these, shooting them from unrevealed positions, and at the estimated positions of the Americans.

Since the Americans, not knowing the exact location of the Han outer line, had shot their barrage over it, and the Hans had fired at unknown American positions, this first exchange of fire had done little more than to churn up vast areas of mountain and valley.

The Hans appeared to be elated, to feel that they had driven off an American attack. I knew better. The next American move, I felt, would be the occupation of the air, from which they had driven the Hans, and from swoopers to direct the rocket fire at the city itself. Then, when they had destroyed this, they would sweep in and hunt down the Hans, man to man, in the surrounding mountains. Command of the air was still important in military strategy, but command of the air rested no longer in the air, but on the ground.

The Hans themselves attempted to scout the American positions from the air, under cover of a massed attack of ships in "cloud bank" or beaming formation, but with very little success. Most of their ships were shot down, and the remainder slid back to the city on sharply inclined repeller rays, one of them which had its generators badly damaged while still fifty miles out, collapsed over the city, before it

CHAPTER XII

could reach its berth at the airport, and crashed down through the glass roof of the city, doing great damage.

Then followed the "air balls," an unforeseen and ingenious resurrection by the Americans of an old principle of air and submarine tactics, through a modern application of the principle of remote control.

The air balls took heavy toll of the morale of the Hans before they were clearly understood by them, and even afterward for that matter.

* * * * *

Their first appearance was quite mysterious. One uneasy night, while the pulsating growl of the distant barrage kept the nerves of the city's inhabitants on edge, there was an explosion near the top of a pinnacle not far from the Imperial Tower. It occurred at the 732nd level, and caused the structure above it to lean and sag, though it did not fall.

Repair men who shot up the shafts a few minutes later to bring new broadcast lamps to replace those which had been shattered, reported what seemed to be a sphere of metal, about three feet in diameter, with a four-inch lens in it, floating slowly down the shaft, as though it were some living creature making a careful examination, pausing now and then as its lens swung about like a great single eye. The moment this "eye" turned upon them, they said, the ball "rushed" down on them, crushing several to death in its vicious gyrations, and jamming the mechanism of the elevator, though failing to crash through it. Then, said the wounded survivors, it floated back up the shaft, watchfully "eyeing" them, and slipped off to the side at the wrecked level.

The next night several of these "air balls" were seen, following explosions in various towers and sections of the city roof and walls. In each case repair gangs were "rushed" by them, and suffered many casualties. On the third night a few of the air balls were destroyed by the repair men and guards, who now were equipped with disintegrator pistols.

This, however, was pretty costly business, for in each case the ray bored into the corridor and shaft walls beyond its target, wrecking much machinery, injuring the structural members of that section, penetrating apartments and taking a number of lives. Moreover, the "air balls," being destroyed, could not be subjected to scientific inspection.

After this the explosions ceased. But for many days the sudden appearances of those "air balls" in the corridors and shafts of the city caused the greatest confusion, and many times they were the cause of death and panic.

At times they released poison gases, and not infrequently themselves burst, instead of withdrawing, in a veritable explosion of disease germs, requiring absolute quarantine by the Han medical department.

There was an utter heartlessness about the defense of the Han authorities, who considered nothing but the good of the community as a whole; for when they established these quarantines, they did not hesitate to seal up thousands of the city's inhabitants behind hermetic barriers enclosing entire sections of different levels, where deprived of food and ventilation, the wretched inhabitants died miserably, long before the disease germs developed in their systems.

* * * * *

At the end of two weeks the entire population of the city was in a mood of panicky revolt. News service to the public had been suspended, and the use of all viewplates and phones in the city were restricted to official communications. The city administration had issued orders that all citizens not on duty should keep to their apartments, but the order was openly flouted, and small mobs were wandering through the corridors, ascending and descending from one level to another, seeking they knew not what, fleeing the air balls, which might appear anywhere, and being driven back from the innermost and deepest sections of the city by the military guard.

I now made up my mind that the time was ripe for me to attempt my escape. In all this confusion I might have an even break, in spite of the danger I might myself run from the air balls, and the almost insuperable difficulties of making my way to the outside of the city and

CHAPTER XII

down the precipitous walls of the mountain to which the city clung like a cap. I would have given much for my inertron belt, that I might simply have leaped outward from the edge of the roof some dark night and floated gently down. I longed for my ultrophone equipment, with which I might have established communication with the beleaguering American forces.

My greatest difficulty, I knew, would be that of escaping my guard. Once free of them, I figured it would be the business of nobody in particular, in that badly disorganized city, to recapture me. The knives of the ordinary citizens I did not fear, and very few of the military guard were armed with disintegrator pistols.

I was sitting in my apartment busying my mind with various plans, when there occurred a commotion in the city corridor outside my door. The captain of my guard jumped nervously from the couch on which he had been reclining, and ordered the excited guards to open the door.

In the broad corridor, the remainder of the guard lay about, dead or groaning, where they had been bowled over by one of these air balls, the first I had ever seen.

The metal sphere floated hesitantly above its victims, turning this way and that to bring its "eye" on various objects around. It stopped dead on sighting the door the guard had thrown open, hesitated a moment, and then shot suddenly into the apartment with a hissing sound, flinging into a far corner one of the guards who had not been quick enough to duck. As the captain drew his disintegrator pistol, it launched itself at him with a vicious hiss. He bounded back from the impact, his chest crushed in, while his pistol, which fortunately had fallen with its muzzle pointed away from me, shot a continuous beam that melted its way instantly through the apartment wall.

* * * * *

The sphere then turned on the other guard, who had thrown himself into a corner where he crouched in fear. Deliberately it seemed to gauge the distance and direction. Then it hurled itself at him with another vicious hiss, which I now saw came from a little rocket motor, crushing him to death where he lay.

It swung slowly around until the lens faced me again, and floated gently into position level with my face, seeming to scan me with its blank, four-inch eye. Then it spoke, with a metallic voice.

"If you are an American," it said, "answer with your name, gang and position."

"I am Anthony Rogers," I replied, still half bewildered, "Boss of the Wyomings. I was captured by the Hans after my swooper was disabled in a fight with a Han airship and had drifted many hundred miles westward. These Hans you have killed were my guard."

"Good!" ejaculated the metal ball. "We have been hunting for you with these remote control rockets for two weeks. We knew you had been captured. A Han message was picked up. Close the door of your room, and hide this ball somewhere. I have turned off the rocket power. Put it on your couch. Throw some pillows over it. Get out of sight. We'll speak softly, so no Hans can hear, and we'll speak only when you speak to us."

The ball, I found, was floating freely in the air. So perfectly was it balanced with ultron and inertron that it had about the weight of a spider web. Ultimately, I suppose, it would have settled to the floor. But I had no time for such an idle experiment. I quickly pushed it to my couch, where I threw a couple of pillows and some of the bed clothes over it. Then I threw myself back on the couch with my head near it. If the dead guards outside attracted attention, and the Han patrol entered, I could report the attack by the "air ball" and claim that I had been knocked unconscious by it.

"One moment," said the ball, after I reported myself ready to talk. "Here is someone who wants to speak to you." And I nearly leaped from the couch with joy when, despite the metallic tone of the instrument, I recognized the eager, loving voice of my wife, almost hysterical in her own joy at talking to me again.

CHAPTER XIII

Escape!

We had little time, however, to waste in endearments, and very little to devote to informing me as to the American plans. The essential thing was that I report the Han plans and resources to the fullest of my ability. And for an hour or two I talked steadily, giving an outline of all I had learned from San-Lan and his Councillors, and particularly of the arrangements for drawing off the population of the city to new cities concealed underground, through the system of tunnels radiating from the base of the mountain. And as a result, the Americans determined to speed up their attack.

There were, as a matter of fact, only two relatively small commands facing the city, Wilma told me, but both of them were picked troops of the new Federal Council. Those to the south were a division of veterans who a few weeks before had destroyed the Han city of Sa-Lus (St. Louis). On the east were a number of the Colorado Gangs and an expeditionary force of our own Wyomings. The attack on Lo-Tan was intended chiefly as an attack on the morale of the Hans of the other twelve cities. If there seemed to be a chance of victory, the operations were to be pushed through. Otherwise the object would be to do as much damage as possible, and fade away into the forests if the Hans developed any real pressure with their new infantry and field batteries of rocket guns and disintegrator-rays.

The "air balls" were simply miniature swoopers of spherical shape, ultronically controlled by operators at control boards miles away, and who saw on their viewplates whatever picture the ultronic television lens in the sphere itself picked up at the predetermined focus. The main propulsive rocket motor was diametrically opposite the lens, so that the sphere could be steered simply by keeping the picture of its objective centered on the crossed hairlines of the viewplates. The outer shell moved magnetically as desired with respect to the core, which was gyroscopically stabilized. Auxiliary rocket motors enabled the operator to make a sphere move sidewise, backward or vertically. Some of these spheres were equipped with devices which enabled their operators to hear as well as see through their ultronic broadcasts,

CHAPTER XIII

and most of those which had invaded the interior of Lo-Tan were equipped with "speakers," in the hope of finding me and establishing communication. Still others were equipped for two-stage control. That is, the operator control led the vision sphere, and through it watched and steered an air torpedo that travelled ahead of it.

The Han airship or any other target selected by the operator of such a combination was doomed. There was no escape. The spheres and torpedoes were too small to be hit. They could travel with the speed of bullets. They could trail a ship indefinitely, hover a safe distance from their mark, and strike at will. Finally, neither darkness nor smoke screens were any bar to their ultronic vision. The spheres, which had penetrated and explored Lo-Tan in their search for me, had floated through breaches in the walls and roofs made by their advance torpedoes.

* * * * *

Wilma had just finished explaining all this to me when I heard a noise outside my door. With a whispered warning I flung myself back on the couch and simulated unconsciousness. When I did not answer the poundings and calls to open, a police detail broke in and shook me roughly.

"The air ball," I moaned, pretending to regain consciousness slowly. "It came in from the corridor. Look what it did to the guard. It must have grazed my head. Where is it?"

"Gone," muttered the under-officer, looking fearfully around. "Yes, undoubtedly gone. These men have been dead some time. And this pistol. The ball got him before he had a chance to use it. See, it has beamed through the wall only here, where he dropped it. Who are you? You look like a tribesman. Oh, yes, you're the Heaven-Born's special prisoner. Maybe I ought to beam you right now. Good thing. Everyone would call it an accident. By the Grand Dragon, I will!"

While he was talking, I had staggered to the other side of the room, to draw his attention away from the couch where the ball was concealed.

CHAPTER XIII

Now suddenly the pillows burst apart, and a blanket with which I had covered the thing streaked from the couch, hitting the man in the small of the back. I could hear his spine snap under the impact. Then it shot through the air toward the group of soldiers in the doorway, bowling them over and sending them shrieking right and left along the corridor. Relentlessly and with amazing speed it launched itself at each in turn, until the corpses lay grotesquely strewn about, and not one had escaped.

It returned to me for all the world like an old-fashioned ghost, the blanket still draped over it (and not interfering with its ultronic vision in the least) and "stood" before me.

"The yellow devils were going to kill you, Tony," I heard Wilma's voice saying. "You've got to get out of there, Tony, before you are killed. Besides, we need you at the control boards, where you can make real use of your knowledge of the city. Have you your jumping belt, ultrophone and rocket gun?"

"No," I replied, "they are all gone."

"It would be no good for you to try to make your way to one of the breaches in the wall, nor to the roof," she mused.

"No, they are too well guarded," I replied, "and even if you made a new one at a predetermined spot I'm afraid the repair men and the patrol would go to it ahead of me."

"Yes, and they would beam you before you could climb inside of a swooper," she added.

"I'll tell you what I can do, Wilma," I suggested. "I know my way about the city pretty well. Suppose I go down one of the shafts to the base of the mountain. I think I can get out. It is dark in the valley, so the Hans cannot see me, and I will stand out in the open, where your ultroscopes can pick me up. Then a swooper can drop quickly down and get me."

CHAPTER XIII

"Good!" Wilma said. "But take that Han's disintegrator pistol with you. And go right away, Tony. But wrap this ball in something and carry it with you. Just toss it from you if you are attacked. I'll stay at the control board and operate it in case of emergency."

* * * * *

So I picked up ball and pistol, and thrust the hand in which I held it into the loose Han blouse I wore, wrapped the ball in a piece of sheeting, and stepped out in the corridor, hurrying toward the nearest magnetic car station, a couple of hundred feet down the corridor, for I had to cross nearly the entire width of the city to reach the shaft that went to the base of the mountain.

I thanked Providence for the perfection of the Han mechanical devices when I reached the station. The automatic checking system of these cars made station attendants unnecessary. I had only to slip the key I had taken from the dead Han officer into the account-charting machine at the station to release a car.

Pressing the proper combinations of main and branch line buttons, I seated myself, holding the pistol ready but concealed beneath my blouse. The car shot with rapid acceleration down the narrow tunnel.

The tubes in which these magnetic cars (which slid along a few inches above the floor of the tunnel by localized repeller rays) ran were very narrow, just the width of the car, and my only danger would come if on catching up to another car its driver should turn around and look in my face. If I kept my face to the front, and hunched over so as to conceal my size, no driver of a following car would suspect that I was not a Han like himself.

The tube dipped under traffic as it came to a trunk line, and my car magnetically lagged, until an opening in the traffic permitted it to swing swiftly into the main line tunnel. At the automatic distance of ten feet it followed a car in which rode a scantily clad girl, her flimsy silks fluttering in the rush of air. I cursed my luck. She would be far more likely to turn around than a man, to see if a man were in the car behind, and if he were personable--for not even the impending doom

CHAPTER XIII

of the city and the public demoralization caused by the "air balls" had dulled the proclivities of the Han women for brazen flirtation. And turn around she did.

Before I could lower my head she had seen my face, and knew I was no Han. I saw her eyebrows arch in surprise. But she seemed puzzled rather than scared. Before she could make up her mind about me, however, her car had swung out of the main tunnel on its predetermined course, and my own automatically was closing up the gap to the car ahead. The passenger in this one wore the uniform of a medical officer, but he did not turn around before I swung out of main traffic to the little station at the head of the shaft.

This particular shaft was intended to serve the very lowest levels exclusively, and since its single car carried nothing but express traffic, it was used only by repair men and other specialists who occasionally had to descend to those levels.

* * * * *

There were only three people on the little platform, which reminded me very much of the subway stations of the Twentieth Century. Two men and a girl stood facing the gate of the shaft, waiting for the car to return from below. One of these was a soldier, apparently off duty, for though he wore the scarlet military coat he carried no weapons other than his knife. The other man wore nothing but sandals and a pair of loose short pants of some heavy and serviceable material. I did not need to look at the compact tool kit and the ray machines attached to his heavy belt, nor the gorgeously jewelled armlet and diadem that he wore to know him for a repair man.

The girl was quite scantily clad, but wore a mask, which was not unusual among the Han women when they went forth on their flirtatious expeditions, and there was something about the sinuous grace of her movements that seemed familiar to me. She was making desperate love to the repair man, whose attitude toward her was that of pleased but lofty tolerance. The soldier, who was seeking no trouble, occupied himself strictly with his own thoughts and paid little attention to them.

CHAPTER XIII

I stepped from my car, still carrying my bundle in which the "air ball" was concealed, and the car shot away as I threw the release lever over. Not so successful as the soldier in simulating lack of interest in the amorous girl and her companion, I drew from the latter a stare of haughty challenge, and the girl herself turned to look at me through her mask.

She gasped as she did so, and shrank back in alarm. And I knew her then in spite of her mask. She was the favorite of the Heaven-Born himself.

"Ngo-Lan!" I exclaimed before I could catch myself.

At the mention of her name, the soldier's head jerked up quickly, and the girl herself gave a little cry of terror, shrinking against her burly companion. This would mean death for her if it reached the ears of her lord.

And her companion, arrogant in his immunity as a repair man, hesitated not a second. His arm shot out toward the soldier, who was nearer to him than I. There was the flash of a knife blade, and the soldier sagged on his feet, then tumbled over like a sack of potatoes, and before my mind had grasped the danger, he had swept the girl aside and was springing at me.

* * * * *

That I lived for a moment even was due to the devotion of my wife, Wilma, who somewhere in the mountains to the east was standing loyally before the control board of the air ball I carried.

For even as the Han leaped at me, the bundle containing the air ball, which I had placed at my feet, shot diagonally upward, catching the fellow in the middle of his leap, hurling him back against the grilled gate of the elevator shaft, and pinning his lifeless body there.

An instant the girl gazed in speechless horror at what had been her secret lover, then she threw herself at my feet, writhing and shrieking in terror.

CHAPTER XIII

At this moment, the elevator shot to a sudden stop behind the grill, and prepared for the worst, I faced it, disintegrator pistol raised.

But I lowered the pistol at once, with a sigh of relief. The elevator was empty. For a moment I considered. I dared not leave either of these bodies nor the girl behind in descending the shaft. At any moment other passengers might glide out of the tunnel to take the elevator, and give an alarm.

So I played the beam of the pistol for an instant on the two dead bodies. They vanished, of course, into nothingness, as did part of the station platform. The damage to the platform, however, would not necessarily be interpreted as evidence of a prisoner escaping.

Then I threw open the elevator gate, dragging Ngo-Lan into the car and stifling her hysterical shrieks, pressed the button that caused it to shoot downward. In a few moments I stepped out several thousand feet below, into a shaft that ran toward one of the Valley Gates.

The pistol again became serviceable, this time for the destruction of the elevator, thus blocking any possible pursuit, yet without revealing my flight.

Ngo-Lan fought like a cat, but despite her writhing, scratching and biting, I bound and gagged her with her own clothing, and left her lying in the tunnel while I stepped in a car and shot toward the gate.

As the car glided swiftly along the brilliantly lit but deserted tunnel I conversed again with Wilma through the metallic speaker of the air ball.

"The only obstacle now," I told her, "is the massive gate at the end of the tunnel. The gate-guard, I think, is posted both outside and inside the gate."

"In that case, Tony," she replied, "I will shoot the ball ahead, and blow out the gate. When you hear it bump against the gate, throw yourself flat in the car, for an instant later I will explode it. Then you can rush through the gate into the night. Scout ships are now hovering above,

CHAPTER XIII

and they will see you with their ultroscopes, though the darkness will leave you invisible to the Hans."

* * * * *

With this the ball shot out of the car and flashed away, down the tunnel ahead of me. I heard a distant metallic thump, and crouched low in the speeding car, clapping my hands to my ears. The heavy detonation which followed, struck me like a blow, and left me gasping for breath. The car staggered like a living thing that had been struck, then gathered speed again and shot forward toward the gaping black hole where the gate had been.

I brought it to a stop at the pile of debris, and climbed through this to freedom and the night. Stumblingly I made my way out into the open, and waited.

Behind, and far above me on the mountain peak, the lights of the city gleamed and flashed, while the iridescent beams of countless disintegrator ray batteries on surrounding mountain peaks, played continuously and nervously, criss-crossing in the sky above it.

Then with a swish, a line dropped out of the sky, and a little seat rested on the ground beside me. I climbed into it, and without further ado was whisked up into the swooper that floated a few hundred feet above me.

A half an hour later I was deposited in a little forest glade where the headquarters of the Wyoming Gang were located, and was greeted with a frantic disregard for decorum by the Deputy Boss of the Wyomings, who rushed upon me like a whirlwind, laughing, crying and whispering endearments all in the same breath, while I squeezed her, Wilma, my wife, until at last she gasped for mercy.

CHAPTER XIV

The Destruction of Lo-Tan

"How did you know I had been taken to Lo-Tan as a prisoner?" I asked the little group of Wyoming Bosses who had assembled in Wilma's tent to greet me. "And how does it happen that our gang is away out here in the Rocky Mountains? I had expected, after the fall of Nu-Yok, that you would join the forest ring around Bah-Flo (Buffalo I called it in the Twentieth Century) or the forces beleaguering Bos-Tan."

They explained that my encounter with the Han airship had been followed carefully by several scopemen. They had seen my swooper shoot skyward out of control, and had followed it with their telultronoscopes until it had been caught in a gale at a high level, and wafted swiftly westward. Ultronophone warnings had been broadcast, asking western Gangs to rescue me if possible. Few of the Gangs west of the Alleghanies, however, had any swoopers, and though I was frequently reported, no attempts could be made to rescue me. Scopemen had reported my capture by the Han ground post, and my probable incarceration in Lo-Tan.

The Rocky Mountain Gangs, in planning their campaign against Lo-Tan, had appealed to the east for help, and Wilma had led the Wyoming veterans westward, though the other eastern Gang had divided their aid between the armies before Bah-Flo and Bos-Tan.

The heavy bombardment which I had heard from Lo-Tan, they told me, was merely a test of the enemy's tactics and strength, but it accomplished little other than to develop that the Hans had the mountains and peaks thickly planted with rocket gunners of their own. It was almost impossible to locate these gun posts, for they were well camouflaged from air observation, and widely scattered; nor did they reveal their positions when they went into action as did their ray batteries.

The Hans apparently were abandoning their rays except for air defense. I told what I knew of the Han plans for abandoning the city, and their escape tunnels. On the strength of this, a general council of

CHAPTER XIV

Gang Bosses was called. This council agreed that immediate action was necessary, for my escape from the city probably would be suspected, and San-Lan would be inclined to start an exodus at once.

* * * * *

As a matter of fact, the destruction of the city presented no real problem to us at all. Explosive air balls could be sent against any target under a control that could not be better were their operators riding within them, and with no risk to the operators. When a ball was exploded on its target by the operator, or destroyed by accident, he simply reported the fact to the supply division, and a fresh one was placed on the jump-off, tuned to his controls.

To my own Gang, the Wyomings, the Council delegated the destruction of the escape tunnels of the enemy. We had a comfortably located camp in a wooded canyon, some hundred and thirty miles northeast of the city, with about 500 men, most of whom were bayonet-gunners, 350 girls as long-gunners and control-board operators, 91 control boards and about 250 five-foot, inertron-protected air balls, of which 200 were of the explosive variety.

I ordered all control boards manned, taking Number One myself, and instructed the others to follow my lead in single file, at the minimum interval of safety, with their projectiles set for signal rather than contact detonation.

In my mind I paid humble tribute to the ingenuity of our engineers as I gently twisted the lever that shot my projectile vertically into the air from the jump-off clearing some half mile away.

The control board before me was a compact contrivance about five feet square. The center of it contained a four-foot viewplate. Whatever view was picked up by the ultronoscope "eye" of the air ball was automatically broadcast on an accurate tuning channel to this viewplate by the automatic mechanism of the projectile. In turn my control board broadcast the signals which automatically controlled the movements of the ball.

CHAPTER XIV

Above and below the viewplate were the pointers and the swinging needles which indicated the speed and angle of vertical movement, the altimeter, the directional compass, and the horizontal speed and distance indicators.

At my left hand was the lever by which I could set the "eye" for penetrative, normal or varying degrees of telescopic vision, and at my right the universally jointed stick (much like the "joy stick" of the ancient airplanes) with its speed control button on the top, with which the ball was directionally "pointed" and controlled.

The manipulation of these levers I had found, with a very little practice, most instinctive and simple.

So, as I have said, I pointed my projectile straight up and let it shoot to the height of two miles. Then I levelled it off, and shot it at full speed (about 500 miles an hour with no allowance for air currents) in a general southwesterly direction, while I eased my controls until I brought in the telescopic view of Lo-Tan. I centered the picture of the city on the crossed hairlines in the middle of my viewpoint, and watched its image grow.

* * * * *

In about fifteen minutes the "string" of air balls was before the city, and speaking in my ultrophone I gave the order to halt, while I swung the scope control to the penetrative setting and let my "eye" rove slowly back and forth through the walls of the city, hunting for a spot from which I might get my bearings. At last, after many penetrations, I managed to bring in a view of the head of the shaft at the bottom of which I knew the tunnels were located, and saw that we were none too soon, for all the corridors leading toward this shaft were packed with Hans waiting their turn to descend.

Slowly I let my "eye" retreat down one of these corridors until I "pulled it out" through the outer wall of the city. There I held the spot on the crossed hairlines and ordered Number Two Operator to my control board, where I pointed out to her the exact spot where I desired a breach in the wall. Returning to her own board, she withdrew her ball

CHAPTER XIV

from the "string," and focussing on this spot in the wall, eased her projectile into contact with it and detonated.

The atomic force of the explosion shattered a vast section of the wall, and for the moment I feared I had balked my own game by not having provided a less powerful projectile.

After some fumbling, however, I was able to maneuver my ball through a gap in the debris and find the corridor I was seeking. Down this corridor I sent it at the speed of a Twentieth Century bullet, (this is to say, about half speed) to spare myself the sight of the slaughter as it cut a swath down the closely packed column of the enemy. If there were any it did not kill, I knew they would be taken care of by the other balls in the string which would follow.

I had to slow it up, however, near the head of the shaft to take my bearings; and a sea of evil faces, contorted with livid terror, looked at me from my viewplate. But not even the terror could conceal the hate in those faces, and there arose in my mind the picture of their long centuries of ruthless cruelty to my race, and the hopelessness of changing the tigerish nature of these Hans. So I steeled myself, and drove the ball again and again into that sea of faces, until I had cleared the station platform of any living enemy, and sent the survivors crushing their way madly along the corridors away from it. There was blinding flash or two on my viewplate as some Han officer tried his ray pistol on my projectile, but that was all, except that he must have disintegrated many of his fellows, for our balls were sheathed in inertron, and suffered no damage themselves.

* * * * *

Cautioning my unit to follow carefully, I pushed my control lever all the way forward until my "eye" pointed down, and there appeared on my viewplate the smooth cylindrical interior of the shaft, fading down toward the base of the mountain, and like a tiny speck, far, far down, was the car, descending with its last load.

I dropped my ball on it, battering it down to the bottom of the shaft, and with hammer-like blows flattening the wreckage, that I might squeeze

CHAPTER XIV

the ball out of the shaft at the lower station.

It emerged into the great vaulted excavation, capable of holding a thousand or more persons, from which the various escape tunnels radiated. Down these tunnels the last remnants of a crowd of fugitives were disappearing, while red-coated soldiers guided the traffic and suppressed disorder with the threat of their spears, and the occasional flourish of a ray pistol.

As I floated my ball out into the middle of the artificial cavern I could see them stagger back in terror. Again the blinding flashes of a few ray pistols, and instantaneous borings of the rays into the walls. The red coats nearest the escape tunnels fled down them in panic. Those whose escape I blocked dropped their weapons and shrank back against the smooth, iridescent green walls.

I marshalled the rest of my string carefully into the cavern, and counted the tunnel entrances, slowly swinging my "eye" around the semicircle of them. There were 26 corridors diverging to the north and west. I decided to send three balls down each, leave 12 in the cavern, then detonate them all at once.

Assigning my operators to their corridors, I ordered intervals of five miles between them, and taking the lead down the first corridor, I ordered "go."

Soon my ball overtook the stream of fugitives, smashing them down despite ray pistols and even rockets that were shot against it. On and on I drove it, time and again battering it through detachments of fleeing Hans, while the distance register on my board climbed to ten, twenty, fifty miles.

Then I called a halt, and suspended my previous orders. I had had no idea that the Hans had bored these tunnels for such distances under the surface of the ground as this. It would be necessary to trace them to their ends and locate their new underground cities in which they expected to establish themselves, and in which many had established themselves by now, no doubt.

CHAPTER XIV

Fifty miles of air in these corridors, I thought, ought to prove a pretty good cushion against the shock of detonation in the cavern. So I ordered detonation of the twelve balls we had left behind. As I expected, there was little effect from it so far out in the tunnels.

But from our scopemen who were covering the city from the outside, I learned that the effects of the explosion on the mountain were terrific; far more than I had dared to hope for.

* * * * *

The mountain itself burst asunder in several spots, throwing out thousands of tons of earth and rock. One-half the city itself tore loose and slid downward, lost in the debris of the avalanche of which it was a part. The remainder, wrenched and convulsed like a living thing in agony, cracked, crumbled and split, towers tumbling down and great fissures appearing in its walls. Its power plant and electro machinery went out of commission. Fifteen of its scout ships hovering in the air directly above, robbed of the power broadcast and their repeller beams disappearing, crashed down into the ruins.

But out in the escape tunnels, we continued our explorations, now sure that no warnings could be broadcast to the tunnel exits, and mowed down contingent after contingent of the hated yellow men.

My register showed seventy-five miles before I came to the end of the tunnel, and drove my ball out into a vast underground city of great, brilliantly illuminated corridors, some of them hundreds of feet high and wide. The architectural scheme was one of lace-like structures of curving lines and of indescribable beauty.

Word had reached us now of the destruction of the city itself, so that no necessity existed for destroying the escape tunnels. In consequence, I ordered the two operators, who were following me, to send their balls out into this underground city, seeking the shaft which the Hans were sure to have as a secret exit to the surface of the earth above.

CHAPTER XIV

But at this juncture events of transcending importance interrupted my plans for a thorough exploration of these new subterranean cities of the Hans. I detonated my projectile at once and ordered all of the operators to do so, and to tune in instantly on new ones. That we wrecked most of these new cities I now know, but of course at the time we were in the dark as to how much damage we caused, since our viewplates naturally went dead when we detonated our projectiles.

CHAPTER XV

The Counter-Attack

The news which caused me to change my plans was grave enough. As I have explained, the American lines lay roughly to the east and the south of the city in the mountains. My own Gang held the northern flank of the east line. To the south of us was the Colorado Union, a force of 5,000 men and about 2,000 girls recruited from about fifteen Gangs. They were a splendid organization, well disciplined and equipped. Their posts, rather widely distributed, occupied the mountain tops and other points of advantage to a distance of about a hundred and fifty miles to the south. There the line turned east, and was held by the Gangs which had come up from the south. Now, simultaneously with the reports from my scouts that a large Han land force was working its way down on us from the north, and threatening to outflank us, came word from Jim Hallwell, Big Boss of the Colorado Union and the commander in chief of our army, that another large Han force was to the southwest of our western flank. And in addition, it seemed, most of the Han military forces at Lo-Tan had been moved out of the city and advanced toward our lines before our air-ball attack.

The situation would not have been in the least alarming if the Hans had had no better arms to fight with than their disintegrator rays, which naturally revealed the locations of their generators the second the visible beams went into play, and their airships, which we had learned how to bring down, first from the air, and now from the ground, through ultrono-controlled projectiles.

But the Hans had learned their lesson from us by this time. Their electrono-chemists had devised atomic projectiles, rocket-propelled, very much like our own, which could be launched in a terrific barrage without revealing the locations of their batteries, and they had equipped their infantry with rocket guns not dissimilar to ours. This division of their army had been expanded by general conscription. So far as ordnance was concerned, we had little advantage over them; although tactically we were still far superior, for our jumping belts enabled our men and girls to scale otherwise inaccessible heights, conceal themselves readily in the upper branches of the giant trees,

CHAPTER XV

and gave them a general all around mobility, the enemy could not hope to equal.

We had the advantage too, in our ultronophones and scopes, in a field of energy which the Hans could not penetrate, while we could cut in on their electrono or (as I would have called it in the Twentieth Century) radio broadcasts.

* * * * *

Later reports showed that there were no less than 10,000 Hans in the force to our north, which evidently was equipped with a portable power broadcast, sufficient for communication purposes and the local operation of small scoutships, painted a green which made them difficult to distinguish against the mountain and forest backgrounds. These ships just skimmed the surface of the terrain, hardly ever outlining themselves against the sky. Moreover, the Han commanders wisely had refrained from massing their forces. They had developed over a very wide and deep front, in small units, well scattered, which were driving down the parallel valleys and canyons like spearheads. Their communications were working well too, for our scouts reported their advance as well restrained, and maintaining a perfect front as between valley and valley, with a secondary line of heavy batteries, moved by small airships from peak to peak, following along the ridges somewhat behind the valley forces.

Hallwell had determined to withdraw our southern wing, pivoting it back to face the outflanking Han force on that side, which had already worked its way well down in back of our line.

In the ultronophone council which we held at once, each Boss tuning in on Hallwell's band, though remaining with his unit, Wilma and I pleaded for a vigorous attack rather than a defensive maneuver. Our suggestion was to divide the American forces into three divisions, with all the swoopers forming a special reserve, and to advance with a rush on the three Han forces behind a rolling barrage.

But the best we could do was to secure permission to make such an attack with our Wyomings, if we wished, to serve as a diversion while

CHAPTER XV

the lines were reforming. And two of the southern Gangs on the west flank, which were eager to get at the enemy, received the same permission.

The rest of the army fumed at the caution of the council, but it spoke well for their discipline that they did not take things in their own hands, for in the eyes of those forest men who had been hounded for centuries, the chance to spring at the throats of the Hans outweighed all other considerations.

So, as the council signed off, Wilma and I turned to the eager faces that surrounded us, and issued our orders.

* * * * *

In a moment the air was filled with leaping figures as the men and girls shot away over the tree tops and up the mountain sides in the deployment movement.

A group of our engineers threw themselves headlong toward a cave across the valley, where they had rigged out a powerful electrono plant operating from atomic energy. And a few moments later the little portable receiver, the Intelligence Boss used to pick up the enemy messages, began to emit such ear-splitting squeals and howls that he shut it off. Our heterodyne or "radio-scrambling" broadcast had gone into operation, emitting impulses of constantly varying wave-length over the full broadcast range and heterodyning the Han communications into futility.

In a little while our scouts came leaping down the valley from the north, and our air balls now were hovering above the Han lines, operators at the control boards near-by painstakingly picking up the pictures of the Han squads struggling down the valleys with their comparatively clumsy weapons.

As fast as the air-ball scopes picked out these squads, their operators, each of whom was in ultronophone communication with a girl long-gunner at some spot in our line, would inform her of the location of the enemy unit, and the latter, after a bit of mathematical calculation,

CHAPTER XV

would send a rocket into the air which would come roaring down on, or very near that unit, and wipe it out.

But for all of that, the number of the Han squads were too much for us. And for every squad we destroyed, fifty advanced.

And though the lines were still several miles apart, in most places, and in some cases with mountain ridges intervening, the Han fire control began to sense the general location of our posts, and things became more serious as their rockets too began to hiss down and explode here and there in our lines, not infrequently killing or maiming one or more of our girls.

The men, our bayonet-gunners, had not as yet suffered, for they were well in advance of the girls, under strict orders to shoot no rockets nor in any way reveal their positions; so the Han rockets were going over their heads.

* * * * *

The Hans in the valleys now were shooting diagonal barrages up the slopes toward the ridges, where they suspected we would be most strongly posted, thus making a cross-fire up the two sides of a ridge, while their heavy batteries, somewhat in the rear, shot straight along the tops of the ridges. But their valley forces were getting out of alignment a bit by now, owing to our heterodyne operations.

I ordered our swoopers, of which we had five, to sweep along above these ridges and destroy the Han batteries.

Up in the higher levels where they were located, the Hans had little cover. A few of their small rep-ray ships rose to meet our swoopers, but were battered down. One swooper they brought to earth with a disintegrator ray beam, by creating a vacuum beneath it, but they did it no serious damage, for its fall was a light one. Subsequently it did tremendous damage, cleaning off an entire ridge.

Another swooper ran into a catastrophe that had one chance in a million of occurring. It hit a heavy Han rocket nose to nose. Inertron

CHAPTER XV

sheathing and all, it was blown into powder.

But the others accomplished their jobs excellently. Small, two-man ships, streaking straight at the Hans at between 600 and 700 miles an hour, they could not be hit except by sheer amazing luck, and they showered their tiny but powerful bombs everywhere as they went.

At the same instant I ordered the girls to cease sharp-shooting, and lay their barrages down in the valleys, with their long-guns set for maximum automatic advance, and to feed the reservoirs as fast as possible, while the bayonet-gunners leaped along close behind this barrage.

Then, with a Twentieth Century urge to see with my own eyes rather than through a viewplate, and to take part in the action, I turned command over to Wilma and leaped away, fifty feet a jump, up the valley, toward the distant flashes and rolling thunder.

CHAPTER XVI

Victory

I had gone five miles, and had paused for a moment, half way up the slope of the valley to get my bearings, when a figure came hurtling through the air from behind, and landed lightly at my side. It was Wilma.

"I put Bill Hearn in command and followed, Tony. I won't let you go into that alone. If you die, I do, too. Now don't argue, dear. I'm determined."

So together we leaped northward again toward the battle. And after a bit we pulled up close behind the barrage.

Great, blinding flashes, like a continuous wall of gigantic fireworks, receded up the valley ahead of us, sweeping ahead of it a seething, tossing mass of debris that seemed composed of all nature, tons of earth, rocks and trees. Ever and anon vast sections of the mountain sides would loosen and slide into the valley.

And, leaping close behind this barrage, with a reckless skill and courage that amazed me, our bayonet-gunners appeared in a continuous series of flashing pictures, outlined in midleap against the wall of fire.

I would not have believed it possible for such a barrage to pass over any of the enemy and leave them unscathed. But it did. For the Hans, operating small disintegrator beams from local or field broadcasts, frantically bored deep, slanting holes in the earth as the fiery tides of explosions rolled up the valleys toward them, and into these probably half of their units were able to throw themselves and escape destruction.

But dazed and staggering they came forth again only to meet death from the terrible, ripping, slashing, cleaving weapons in the hands of our leaping bayonet-gunners.

CHAPTER XVI

Thrust! Cut! Crunch! Slice! Thrust! Up and down with vicious, tireless, flashing speed, swung the bayonets and ax-bladed butts of the American gunners as they leaped and dodged, ever forward, toward new opponents.

Weakly and ineffectually the red-coated Han soldiery thrust at them with spears, flailing with their short-swords and knives, or whipping about their ray pistols. The forest men were too powerful, too fast in their remorselessly efficient movement.

With a shout of unholy joy, I gripped a bayonet-gun from the hands of a gunner whose leg had been whisked out of existence beneath him by a pistol ray, and leaped forward into the fight, launching myself at a red-coated officer who was just stepping out of a "worm hole."

Like a shriek of the Valkyrie, Wilma's battle cry rang in my ear as she, too, shot herself like a rocket at a red-coated figure.

I thrust with every ounce of my strength. The Han officer, grinning wickedly as he tried to raise the muzzle of his pistol, threw himself backward as my bayonet ripped the air under his nose. But his grin turned instantly to sickened surprise as the up-cleaving ax-blade on the butt of my weapon caught him in the groin, half bisecting him.

And from the corner of my eye I saw Wilma bury her bayonet in her opponent, screaming in ecstatic joy.

* * * * *

And so, in a matter of seconds, we found ourselves in the front rank, thrusting, cutting, dodging, leaping along behind that blinding and deafening barrage in a veritable whirlwind of fury, until it seemed to me that we were exulting in a consciousness of excelling even that tide of destruction in our merciless efficiency.

At last we became aware, in but a vague sort of way at first, that no more red-coats were rising up out of the ground to go down again before our whirling, swinging weapons. Gradually we paused, looking about in wonder. Then the barrage ceased, and the sudden absence

CHAPTER XVI

of the deafening roll, and the wall of light, in themselves, deafened and blinded us.

I leaped weakly toward the spot where hazily I spied Wilma, now drooping and swaying on her feet, supported as she was by her jumping belt, and caught her in my arms, just as she was sinking gently to the ground.

All around us the weary warriors, crimsoned now with the blood of the enemy, were sinking to the ground in exhaustion. And as I too, sank down, clutching in my arms the unconscious form of my warrior wife, I began to hear, through my helmet phones, the exultant report of headquarters.

Our attack had swept straight through the enemy's sector, completely annihilating everything except a few hundred of his troops on either flank. And these, in panic and terror, had scattered wildly in flight. We had wiped out a force more than ten times our own number. The right flank of the American army was saved. And already the Colorado Union, from behind us, was leaping around in a great circling movement, closing in on the Han force that was advancing from the ruins of Lo-Tan.

Far away, to the southwest, the southern Gangs, reinforced in the end by the bulk of our left wing, had struck straight at the enveloping Han force shattering it like a thunderbolt, and at present were busily hunting down and destroying its scattered remnants.

But before the Colorado Union could complete the destruction of the central division of the enemy, the despairing Hans saved them the trouble. Company after company of them, knowing no escape was possible, lined up in the forest glades and valleys, while their officers swept them out of existence by the hundreds with their ray pistols, which they then turned on themselves.

And so the fall of Lo-Tan was accomplished. Somewhere in the seething activities of these few days, San-Lan, the "Heaven-Born," Emperor of the Hans in America, perished, for he was heard of never again, and the unified action of the Hans vanished with him, though it was several years before one by one their remaining cities were destroyed and their populations hunted down, thus completing the reclamation of America and inaugurating the most glorious and noble era of scientific civilization in the history of the American race.

* * * * *

As I look back on those emotional and violent years from my present vantage point of declining existence in an age of peace and good will toward all mankind, they do seem savage and repellent.

Then there flashes into my memory the picture of Wilma (now long since gone to her rest) as, screaming in an utter abandon of merciless fury, she threw herself recklessly, exultantly into the thick of that wild, relentless slaughter; and my mind can find nothing savage nor repellent about her.

If I, product of the relatively peaceful Twentieth Century, was so completely carried away by the fury of that war, intensified by centuries of unspeakable cruelty on the part of the yellow men who were mentally gods and morally beasts, shall I be shocked at the "bloodthirstiness" of a mate who was, after all, but a normal girl of that day, and who, girl as she was, never for a moment faltered in the high courage with which she threw herself into that combat, responding to the passionate urge for freedom in her blood that not five centuries of inhuman persecution could subdue?

Had the Hans been raging tigers, or slimy, loathsome reptiles, would we have spared them? And when in their centuries of degradation they had destroyed the souls within themselves, were they in any way superior to tigers or snakes? To have extended mercy would have been suicide.

In the years that followed, Wilma and I travelled nearly every nation on the earth which had succeeded in throwing off the Han domination, spurred on by our success in America, and I never knew her to show to the men or women of any race anything but the utmost of sympathetic courtesy and consideration, whether they were the noble brown-skinned Caucasians of India, the sturdy Balkanites of Southern

CHAPTER XVI

Europe, or the simple, spiritual Blacks of Africa, today one of the leading races of the world, although in the Twentieth Century we regarded them as inferior. This charity and gentleness of hers did not fail even in our contacts with the non-Han Mongolians of Japan and the coast provinces of China.

But that monstrosity among the races of men which originated as a hybrid somewhere in the dark fastnesses of interior Asia, and spread itself like an inhuman yellow blight over the face of the globe--for that race, like all of us, she felt nothing but horror and the irresistible urge to extermination.

* * * * *

Latterly, our historians and anthropologists find much support for the theory that the Hans sprang from a genus of human-like creatures that may have arrived on this earth with a small planet (or large meteor) which is known to have crashed in interior Asia late in the Twentieth Century, causing certain permanent changes in the earth's orbit and climate.

Geological convulsions blocked this section off from the rest of the world for many years. And it is a historical fact that Chinese scientists, driving their explorations into it at a somewhat later period, met the first wave of the on-coming Hans.

The theory is that these creatures (and certain queer skeletons have been found in the "Asiatic Bowl") with a mental superdevelopment, but a vacuum in place of that intangible something we call a soul, mated forcibly with the Tibetans, thereby strengthening their physical structure to almost the human normal, adapting themselves to earthly speech and habits, and in some strange manner intensifying even further their mental powers.

Or, to put it the other way around. These Tibetans, through the injection of this unearthly blood, deteriorated slightly physically, lost the "soul" parts of their nature entirely, and developed abnormally efficient intellects.

CHAPTER XVI

However, through the centuries that followed, as the Hans spread over the face of the earth, this unearthly strain in them not only became more dilute, but lost its potency; and in the end, the poison of it submerged the power of it, and earth's mankind came again into possession of its inheritance.

How all this may be, I do not know. It is merely a hypothesis over which the learned men of today quarrel.

* * * * *

But I do know that there was something inhuman about these Hans. And I had many months of intimate contact with them, and with their Emperor in America. I can vouch for the fact that even in his most friendly and human moments, there was an inhumanity, or perhaps "unhumanity" about him that aroused in me that urge to kill.

But whether or not there was in these people blood from outside this planet, the fact remains that they have been exterminated, that a truly human civilization reigns once more--and that I am now a very tired old man, waiting with no regrets for the call which will take me to another existence.

There, it is my hope and my conviction that my courageous mate of those bloody days waits for me with loving arms.

THE END

Transcriber's Note:

In this text the two prefixes *ultro-* and *ultrono-* have been applied inconsistently to much of the future technology. These discrepancies remain as printed.

End of Project Gutenberg's The Airlords of Han, by Philip Francis Nowlan

*** END OF THIS PROJECT GUTENBERG EBOOK THE AIRLORDS OF HAN ***

***** This file should be named 25438.txt or 25438.zip ***** This and all associated files of various formats will be found in:
http://www.gutenberg.org/2/5/4/3/25438/

Produced by Greg Weeks, Stephen Blundell and the Online Distributed Proofreading Team at http://www.pgdp.net

Updated editions will replace the previous one--the old editions will be renamed.

Creating the works from public domain print editions means that no one owns a United States copyright in these works, so the Foundation (and you!) can copy and distribute it in the United States without permission and without paying copyright royalties. Special rules, set forth in the General Terms of Use part of this license, apply to copying and distributing Project Gutenberg-tm electronic works to protect the PROJECT GUTENBERG-tm concept and trademark. Project Gutenberg is a registered trademark, and may not be used if you charge for the eBooks, unless you receive specific permission. If you do not charge anything for copies of this eBook, complying with the rules is very easy. You may use this eBook for nearly any purpose such as creation of derivative works, reports, performances and research. They may be modified and printed and given away--you may do practically ANYTHING with public domain eBooks. Redistribution is subject to the trademark license, especially commercial redistribution.

*** START: FULL LICENSE ***

THE FULL PROJECT GUTENBERG LICENSE PLEASE READ THIS BEFORE YOU DISTRIBUTE OR USE THIS WORK

To protect the Project Gutenberg-tm mission of promoting the free distribution of electronic works, by using or distributing this work (or any other work associated in any way with the phrase "Project Gutenberg"), you agree to comply with all the terms of the Full Project Gutenberg-tm License (available with this file or online at http://gutenberg.net/license).

CHAPTER XVI

Section 1. General Terms of Use and Redistributing Project Gutenberg-tm electronic works

1.A. By reading or using any part of this Project Gutenberg-tm electronic work, you indicate that you have read, understand, agree to and accept all the terms of this license and intellectual property (trademark/copyright) agreement. If you do not agree to abide by all the terms of this agreement, you must cease using and return or destroy all copies of Project Gutenberg-tm electronic works in your possession. If you paid a fee for obtaining a copy of or access to a Project Gutenberg-tm electronic work and you do not agree to be bound by the terms of this agreement, you may obtain a refund from the person or entity to whom you paid the fee as set forth in paragraph 1.E.8.

1.B. "Project Gutenberg" is a registered trademark. It may only be used on or associated in any way with an electronic work by people who agree to be bound by the terms of this agreement. There are a few things that you can do with most Project Gutenberg-tm electronic works even without complying with the full terms of this agreement. See paragraph 1.C below. There are a lot of things you can do with Project Gutenberg-tm electronic works if you follow the terms of this agreement and help preserve free future access to Project Gutenberg-tm electronic works. See paragraph 1.E below.

1.C. The Project Gutenberg Literary Archive Foundation ("the Foundation" or PGLAF), owns a compilation copyright in the collection of Project Gutenberg-tm electronic works. Nearly all the individual works in the collection are in the public domain in the United States. If an individual work is in the public domain in the United States and you are located in the United States, we do not claim a right to prevent you from copying, distributing, performing, displaying or creating derivative works based on the work as long as all references to Project Gutenberg are removed. Of course, we hope that you will support the Project Gutenberg-tm mission of promoting free access to electronic works by freely sharing Project Gutenberg-tm works in compliance with the terms of this agreement for keeping the Project Gutenberg-tm name associated with the work. You can easily comply with the terms of this agreement by keeping this work in the same format with its

CHAPTER XVI

attached full Project Gutenberg-tm License when you share it without charge with others.

1.D. The copyright laws of the place where you are located also govern what you can do with this work. Copyright laws in most countries are in a constant state of change. If you are outside the United States, check the laws of your country in addition to the terms of this agreement before downloading, copying, displaying, performing, distributing or creating derivative works based on this work or any other Project Gutenberg-tm work. The Foundation makes no representations concerning the copyright status of any work in any country outside the United States.

1.E. Unless you have removed all references to Project Gutenberg:

1.E.1. The following sentence, with active links to, or other immediate access to, the full Project Gutenberg-tm License must appear prominently whenever any copy of a Project Gutenberg-tm work (any work on which the phrase "Project Gutenberg" appears, or with which the phrase "Project Gutenberg" is associated) is accessed, displayed, performed, viewed, copied or distributed:

This eBook is for the use of anyone anywhere at no cost and with almost no restrictions whatsoever. You may copy it, give it away or re-use it under the terms of the Project Gutenberg License included with this eBook or online at www.gutenberg.net

1.E.2. If an individual Project Gutenberg-tm electronic work is derived from the public domain (does not contain a notice indicating that it is posted with permission of the copyright holder), the work can be copied and distributed to anyone in the United States without paying any fees or charges. If you are redistributing or providing access to a work with the phrase "Project Gutenberg" associated with or appearing on the work, you must comply either with the requirements of paragraphs 1.E.1 through 1.E.7 or obtain permission for the use of the work and the Project Gutenberg-tm trademark as set forth in paragraphs 1.E.8 or 1.E.9.

1.E.3. If an individual Project Gutenberg-tm electronic work is posted with the permission of the copyright holder, your use and distribution must comply with both paragraphs 1.E.1 through 1.E.7 and any additional terms imposed by the copyright holder. Additional terms will be linked to the Project Gutenberg-tm License for all works posted with the permission of the copyright holder found at the beginning of this work.

1.E.4. Do not unlink or detach or remove the full Project Gutenberg-tm License terms from this work, or any files containing a part of this work or any other work associated with Project Gutenberg-tm.

1.E.5. Do not copy, display, perform, distribute or redistribute this electronic work, or any part of this electronic work, without prominently displaying the sentence set forth in paragraph 1.E.1 with active links or immediate access to the full terms of the Project Gutenberg-tm License.

1.E.6. You may convert to and distribute this work in any binary, compressed, marked up, nonproprietary or proprietary form, including any word processing or hypertext form. However, if you provide access to or distribute copies of a Project Gutenberg-tm work in a format other than "Plain Vanilla ASCII" or other format used in the official version posted on the official Project Gutenberg-tm web site (www.gutenberg.net), you must, at no additional cost, fee or expense to the user, provide a copy, a means of exporting a copy, or a means of obtaining a copy upon request, of the work in its original "Plain Vanilla ASCII" or other form. Any alternate format must include the full Project Gutenberg-tm License as specified in paragraph 1.E.1.

1.E.7. Do not charge a fee for access to, viewing, displaying, performing, copying or distributing any Project Gutenberg-tm works unless you comply with paragraph 1.E.8 or 1.E.9.

1.E.8. You may charge a reasonable fee for copies of or providing access to or distributing Project Gutenberg-tm electronic works provided that

CHAPTER XVI

- You pay a royalty fee of 20% of the gross profits you derive from the use of Project Gutenberg-tm works calculated using the method you already use to calculate your applicable taxes. The fee is owed to the owner of the Project Gutenberg-tm trademark, but he has agreed to donate royalties under this paragraph to the Project Gutenberg Literary Archive Foundation. Royalty payments must be paid within 60 days following each date on which you prepare (or are legally required to prepare) your periodic tax returns. Royalty payments should be clearly marked as such and sent to the Project Gutenberg Literary Archive Foundation at the address specified in Section 4, "Information about donations to the Project Gutenberg Literary Archive Foundation."

- You provide a full refund of any money paid by a user who notifies you in writing (or by e-mail) within 30 days of receipt that s/he does not agree to the terms of the full Project Gutenberg-tm License. You must require such a user to return or destroy all copies of the works possessed in a physical medium and discontinue all use of and all access to other copies of Project Gutenberg-tm works.

- You provide, in accordance with paragraph 1.F.3, a full refund of any money paid for a work or a replacement copy, if a defect in the electronic work is discovered and reported to you within 90 days of receipt of the work.

- You comply with all other terms of this agreement for free distribution of Project Gutenberg-tm works.

1.E.9. If you wish to charge a fee or distribute a Project Gutenberg-tm electronic work or group of works on different terms than are set forth in this agreement, you must obtain permission in writing from both the Project Gutenberg Literary Archive Foundation and Michael Hart, the owner of the Project Gutenberg-tm trademark. Contact the Foundation as set forth in Section 3 below.

1.F.

1.F.1. Project Gutenberg volunteers and employees expend considerable effort to identify, do copyright research on, transcribe and proofread public domain works in creating the Project Gutenberg-tm

CHAPTER XVI

collection. Despite these efforts, Project Gutenberg-tm electronic works, and the medium on which they may be stored, may contain "Defects," such as, but not limited to, incomplete, inaccurate or corrupt data, transcription errors, a copyright or other intellectual property infringement, a defective or damaged disk or other medium, a computer virus, or computer codes that damage or cannot be read by your equipment.

1.F.2. LIMITED WARRANTY, DISCLAIMER OF DAMAGES - Except for the "Right of Replacement or Refund" described in paragraph 1.F.3, the Project Gutenberg Literary Archive Foundation, the owner of the Project Gutenberg-tm trademark, and any other party distributing a Project Gutenberg-tm electronic work under this agreement, disclaim all liability to you for damages, costs and expenses, including legal fees. YOU AGREE THAT YOU HAVE NO REMEDIES FOR NEGLIGENCE, STRICT LIABILITY, BREACH OF WARRANTY OR BREACH OF CONTRACT EXCEPT THOSE PROVIDED IN PARAGRAPH F3. YOU AGREE THAT THE FOUNDATION, THE TRADEMARK OWNER, AND ANY DISTRIBUTOR UNDER THIS AGREEMENT WILL NOT BE LIABLE TO YOU FOR ACTUAL, DIRECT, INDIRECT, CONSEQUENTIAL, PUNITIVE OR INCIDENTAL DAMAGES EVEN IF YOU GIVE NOTICE OF THE POSSIBILITY OF SUCH DAMAGE.

1.F.3. LIMITED RIGHT OF REPLACEMENT OR REFUND - If you discover a defect in this electronic work within 90 days of receiving it, you can receive a refund of the money (if any) you paid for it by sending a written explanation to the person you received the work from. If you received the work on a physical medium, you must return the medium with your written explanation. The person or entity that provided you with the defective work may elect to provide a replacement copy in lieu of a refund. If you received the work electronically, the person or entity providing it to you may choose to give you a second opportunity to receive the work electronically in lieu of a refund. If the second copy is also defective, you may demand a refund in writing without further opportunities to fix the problem.

1.F.4. Except for the limited right of replacement or refund set forth in paragraph 1.F.3, this work is provided to you 'AS-IS' WITH NO OTHER

CHAPTER XVI

WARRANTIES OF ANY KIND, EXPRESS OR IMPLIED, INCLUDING BUT NOT LIMITED TO WARRANTIES OF MERCHANTIBILITY OR FITNESS FOR ANY PURPOSE.

1.F.5. Some states do not allow disclaimers of certain implied warranties or the exclusion or limitation of certain types of damages. If any disclaimer or limitation set forth in this agreement violates the law of the state applicable to this agreement, the agreement shall be interpreted to make the maximum disclaimer or limitation permitted by the applicable state law. The invalidity or unenforceability of any provision of this agreement shall not void the remaining provisions.

1.F.6. **INDEMNITY**

- You agree to indemnify and hold the Foundation, the trademark owner, any agent or employee of the Foundation, anyone providing copies of Project Gutenberg-tm electronic works in accordance with this agreement, and any volunteers associated with the production, promotion and distribution of Project Gutenberg-tm electronic works, harmless from all liability, costs and expenses, including legal fees, that arise directly or indirectly from any of the following which you do or cause to occur: (a) distribution of this or any Project Gutenberg-tm work, (b) alteration, modification, or additions or deletions to any Project Gutenberg-tm work, and (c) any Defect you cause.

Section 2. Information about the Mission of Project Gutenberg-tm

Project Gutenberg-tm is synonymous with the free distribution of electronic works in formats readable by the widest variety of computers including obsolete, old, middle-aged and new computers. It exists because of the efforts of hundreds of volunteers and donations from people in all walks of life.

Volunteers and financial support to provide volunteers with the assistance they need, is critical to reaching Project Gutenberg-tm's goals and ensuring that the Project Gutenberg-tm collection will remain freely available for generations to come. In 2001, the Project Gutenberg Literary Archive Foundation was created to provide a secure and permanent future for Project Gutenberg-tm and future

CHAPTER XVI

generations. To learn more about the Project Gutenberg Literary Archive Foundation and how your efforts and donations can help, see Sections 3 and 4 and the Foundation web page at http://www.pglaf.org.

Section 3. Information about the Project Gutenberg Literary Archive Foundation

The Project Gutenberg Literary Archive Foundation is a non profit 501(c)(3) educational corporation organized under the laws of the state of Mississippi and granted tax exempt status by the Internal Revenue Service. The Foundation's EIN or federal tax identification number is 64-6221541. Its 501(c)(3) letter is posted at http://pglaf.org/fundraising. Contributions to the Project Gutenberg Literary Archive Foundation are tax deductible to the full extent permitted by U.S. federal laws and your state's laws.

The Foundation's principal office is located at 4557 Melan Dr. S. Fairbanks, AK, 99712., but its volunteers and employees are scattered throughout numerous locations. Its business office is located at 809 North 1500 West, Salt Lake City, UT 84116, (801) 596-1887, email business@pglaf.org. Email contact links and up to date contact information can be found at the Foundation's web site and official page at http://pglaf.org

For additional contact information: Dr. Gregory B. Newby Chief Executive and Director gbnewby@pglaf.org

Section 4. Information about Donations to the Project Gutenberg Literary Archive Foundation

Project Gutenberg-tm depends upon and cannot survive without wide spread public support and donations to carry out its mission of increasing the number of public domain and licensed works that can be freely distributed in machine readable form accessible by the widest array of equipment including outdated equipment. Many small donations ($1 to $5,000) are particularly important to maintaining tax exempt status with the IRS.

CHAPTER XVI

The Foundation is committed to complying with the laws regulating charities and charitable donations in all 50 states of the United States. Compliance requirements are not uniform and it takes a considerable effort, much paperwork and many fees to meet and keep up with these requirements. We do not solicit donations in locations where we have not received written confirmation of compliance. To SEND DONATIONS or determine the status of compliance for any particular state visit http://pglaf.org

While we cannot and do not solicit contributions from states where we have not met the solicitation requirements, we know of no prohibition against accepting unsolicited donations from donors in such states who approach us with offers to donate.

International donations are gratefully accepted, but we cannot make any statements concerning tax treatment of donations received from outside the United States. U.S. laws alone swamp our small staff.

Please check the Project Gutenberg Web pages for current donation methods and addresses. Donations are accepted in a number of other ways including including checks, online payments and credit card donations. To donate, please visit: http://pglaf.org/donate

Section 5. General Information About Project Gutenberg-tm electronic works.

Professor Michael S. Hart is the originator of the Project Gutenberg-tm concept of a library of electronic works that could be freely shared with anyone. For thirty years, he produced and distributed Project Gutenberg-tm eBooks with only a loose network of volunteer support.

Project Gutenberg-tm eBooks are often created from several printed editions, all of which are confirmed as Public Domain in the U.S. unless a copyright notice is included. Thus, we do not necessarily keep eBooks in compliance with any particular paper edition.

Most people start at our Web site which has the main PG search facility:

http://www.gutenberg.net

This Web site includes information about Project Gutenberg-tm, including how to make donations to the Project Gutenberg Literary Archive Foundation, how to help produce our new eBooks, and how to subscribe to our email newsletter to hear about new eBooks.

The Airlords of Han, by Philip Francis Nowlan

A free ebook from http://manybooks.net/

www.ingramcontent.com/pod-product-compliance
Lightning Source LLC
Chambersburg PA
CBHW050114230526
45470CB00004B/1822